Jacqueline Cochran

JACQUELINE COCHRAN
Biography of a Pioneer Aviator

Rhonda Smith-Daugherty

McFarland & Company, Inc., Publishers
Jefferson, North Carolina, and London

LIBRARY OF CONGRESS CATALOGUING-IN-PUBLICATION DATA

Smith-Daugherty, Rhonda, 1967–
    Jacqueline Cochran : biography of a pioneer aviator / Rhonda Smith-Daugherty.
        p.   cm.
    Includes bibliographical references and index.

    ISBN 978-0-7864-6275-9
    softcover : acid free paper ∞

    1. Cochran, Jacqueline.   2. Women Airforce Service Pilots (U.S.)—Biography.   3. Air pilots, Military—United States—Biography.   4. Women air pilots—United States—Biography.   5. World War, 1939–1945—Aerial operations, American.   I. Title.
D790.5.S65 2012
940.54'4973092—dc23
[B]   2012007133

BRITISH LIBRARY CATALOGUING DATA ARE AVAILABLE

© 2012 Rhonda Smith-Daugherty. All rights reserved

No part of this book may be reproduced or transmitted in any form or by any means, electronic or mechanical, including photocopying or recording, or by any information storage and retrieval system, without permission in writing from the publisher.

On the cover: Jacqueline Cochran on the cockpit sill of a North American P-51B Mustang, 1946 (Smithsonian Air and Space Museum)

Front cover design by Rob Russell

Manufactured in the United States of America

McFarland & Company, Inc., Publishers
    Box 611, Jefferson, North Carolina 28640
        www.mcfarlandpub.com

For my husband,
Dr. Leo J. Daugherty, III

Table of Contents

Acknowledgments ix
Introduction 1

1. The Making of Jacqueline Cochran 9
2. The Lady Is a Pilot 20
3. America's Number One Woman Flier 35
4. An Unusual Life 54
5. Women in War 65
6. Jacqueline Cochran and the WASPs 80
7. The WASP Experience 98
8. The End of the WASPs 117
9. High Flying 132
10. Postscript 151

Appendix One: Biographical Sketches
- *General Henry Harley Arnold, 1886–1950* 155
- *Vincent Hugo Bendix, 1882–1945* 158
- *Amelia Earhart, 1897–1937* 159
- *Colonel Oveta Culp Hobby, 1905–1995* 161
- *Nancy Harkness Love, 1914–1976* 162
- *General of the Army George Catlett Marshall, 1880–1959* ... 164
- *Floyd Odlum, 1892–1976* 166
- *Alexander Nikolaivich Prokofieff de Seversky, 1894–1974* .. 167
- *Lieutenant General Barton Kyle Yount, 1884–1949* 169

Table of Contents

Appendix Two: Major Aircraft Flown by Jacqueline Cochran 172
Chapter Notes 173
Bibliography 193
Index 205

Acknowledgments

The following book, derived primarily from my dissertation, while an individual work, benefited from the insights and direction of several people. I wish to thank my dissertation chair, Dr. David Hamilton, my other committee members and the outside reader for allowing me to complete this project on schedule. I wish to thank the archivists at the Eisenhower Presidential Library, Air and Space Museum and National Archives for their help. I would also like to thank Archie Difante of Maxwell Air Force Base for his kindness and encouragement. Thanks also go out to the assistance of the staff at the U.S. Air Force's Museum at Wright Patterson AFB, Dayton, Ohio, who were most helpful in facilitating my research.

My thanks go to the staffs at the University of Louisville's Ekstrom and Law School Libraries, and in particular to Chris Poche, Robin Harris, Scott Campbell, and Jerome Neukarch for their technical help and much-needed patience in locating (among other sources) Jacqueline Cochran's testimony before Congress in 1962. They went out of their way to assist me with my research, and for that I am ever grateful.

In addition to the technical and instrumental assistance above, I received equally important assistance from family, friends and colleagues. The faculty at Lexington Community College and Eastern Kentucky University provided much-needed encouragement and support. I am particularly grateful to the staff of the History Department at Western Kentucky University for their unfailing faith and confidence in me, and to Dr. Jack Thacker who never doubted that I would finish. I would also like to thank my parents, Charles and the late Sarah Jane Smith, and my best friend, Gina Sims, for their support throughout the years.

Several people helped me prepare this manuscript for publication.

Acknowledgments

I wish to thank my colleague Dr. Paul Beasely of the History Department at Alice Lloyd College, for his friendship and mentoring. I am also grateful to my assistants for their hard work and dedication — Belinda Boyd, Chantel Gray, Patrick Hall, Stephanie Kilburn, and Leela Thomas, all of whom spent many hours retyping this manuscript and helping with the index. I am especially grateful to my husband, Dr. Leo J. Daugherty III, who encouraged me to publish this manuscript. He also provided much help with the formatting, procurement and arrangement of photographs and other editorial requirements.

Introduction

More than 25 years ago, women pilots began flying for the American military. In 1973, six women became the Navy's first female aviators. A year later, the Army began training female helicopter pilots. Ironically, the Air Force was the last branch to allow women to fly, admitting women to the pilot training program in 1976. Today, about 300 women fly for the Air Force; another 225 for the Navy, and another 567 fly for the Army. These are not the first women, however, to fly military aircraft. During World War II, over 1,000 women flew fighters and bombers in support of the United States Army Air Corps. These women were members of the Women's Air Service Pilots, more commonly known as the WASPs.

The WASPs was a civilian organization, created and directed by the famous aviator Jacqueline Cochran. Although barred from combat, the WASPs still made an important contribution to the war effort. They delivered military aircraft, tested airplanes, towed targets for artillery practice, and released hundreds of male pilots for combat duty.

The WASPs existed largely for two reasons: the great need for pilots in 1942–1943, and the hard work of Jacqueline Cochran. To aid the war effort, Cochran recruited, trained and supervised the WASPs. According to former WASP Dora Dougherty Strother, "[Cochran] fought for us. She fought for the concept that she believed in, that women could do this. She probably was the only person that could have done it as effectively."[1]

Jacqueline Cochran was the logical choice to head an organization of women pilots. While Amelia Earhart was the best-known woman pilot of the 1930s, Cochran was the more important aviation pioneer. She was, in fact, America's top woman pilot. On the eve of World War II, her

Introduction

skill as a pilot had brought her great fame and had transformed her into an aviation icon. This work concentrates on Jacqueline Cochran's early career as a pilot and her role in creating and leading the WASPs during the war. Cochran's postwar aviation career is also discussed, including her records in the new jet aircraft and participation in testing astronauts for the space program. Other aspects of Cochran are explored — her 1956 bid for Congress and her continued reluctance to crusade for advancement for women.

* * *

Jacqueline Cochran began flying in 1932, and she gained national fame as a pilot when she beat nine men in the 1938 Bendix Air Race. According to one article: "Jacqueline Cochran in a short period of time has gained the reputation as one of the outstanding women flyers in the world ... [and] has proved that she can fly the fastest ships, the most powerful engines, and intricate equipment with the best of them."[2] She flew as a barnstormer, tested new airplanes, and entered air races. She established several speed and altitude records and accomplished many "firsts" for women in aviation. In 1937 alone, she established three national records, a new world's women's speed record of 293.06 M.P.H. and flew nonstop from New York to Miami in four hours, breaking all speed records between these cities. Two years later, she established the women's national altitude record and broke the international open-class speed record for men and women. She was the first woman to make a blind landing, pilot an airplane over the Atlantic, and fly military aircraft. Cochran was also the first woman to break the sonic barrier (flying twice the speed of sound), and the first to take off and land a jet on an aircraft carrier. By the end of her life, she had won 15 Clifford Burke Harmon International Trophies of the International League of Aviators as the outstanding woman flyer in the world.

Cochran's position as an internationally known flyer and wife of millionaire Floyd Odlum made possible her role in World War II. When the war began in 1939, Cochran anticipated America's entrance and began campaigning for a new role for women pilots in the military. In 1941, she organized a small group of American women pilots who ferried airplanes for England's Royal Air Force. The following year, she began

Introduction

recruiting women for an American women's pilot program and, in 1942, she was made director of the Women's Air Force Service Pilots (WASPs).

The WASP program was made possible by Cochran's dogged persistence. As early as 1939, she had written Eleanor Roosevelt urging that women be allowed to fly for the military. After America entered the war in 1941, Cochran was chosen to recruit and train women pilots for American service. As the WASP's chief administrator, she sent out questionnaires, interviewed applicants, chose the training facility, found instructors and airplanes and mandated types of training.

This persistence was a quality Cochran had displayed throughout her life as she fought to escape poverty, learn to fly and achieve economic independence. Cochran was a self-made woman with a genuine "rags-to-riches" story. She began her life as a poor orphan; she ended it as a wealthy, internationally known celebrity. Born into poverty in the early 1900s, she transformed herself into a glamorous woman, married a millionaire businessman, and started her own cosmetics company, Jacqueline Cochran Cosmetics. "People used to call her the Golden Girl," explained Cochran's secretary, "because she was always sort of golden."[3] A tall, striking blonde with large brown eyes and creamy skin, Cochran was noticeably feminine.

Cochran was also a gutsy, outspoken individual, determined to do things her way. According to pilot Chuck Yeager, "She was tough and bossy and used to getting her own way. When Jackie Cochran set her mind to do something, she was a damned Sherman tank at full steam."[4] "Jacqueline Cochran," commented Robert Arnold (the grandson of General Henry Arnold), "had tenacity of purpose."[5]

Cochran also had a flare for danger. She flew in acrobatic air shows, tested experimental aircraft and raced unfinished planes, seemingly oblivious to death. During World War II, she flew a bomber across the Atlantic Ocean and lived in wartime London.

Throughout her life, but especially in the 1930s and 1940s, Cochran sought personal independence and desired fame. She literally chose her own name (from a phone book) and remained "Miss Cochran" after her marriage. Despite almost no formal education, she started a successful cosmetics business in 1935. As an aviator, she found a different kind of freedom — the freedom and independence of the sky. Usually, she flew alone.

Introduction

Both in appearance and personal beliefs, Cochran was very different from her friend Amelia Earhart, who sported short cropped hair, wore slacks and used no make-up. Earhart, as Susan Ware has described her, was a "liberal feminist," who sought to advance women's issues.[6] Jacqueline Cochran sought to advance herself and she did little to advance the cause of women. A traditionalist and conformist on most social issues, Cochran held the views of her rural Southern roots. She disliked feminism and insisted publicly that women should be good wives and mothers.

Yet Cochran's actions and career often conflicted with her attitudes. As a pilot, she both challenged and perpetuated gender stereotypes. As a daring aviator, Cochran was an example of what women could accomplish in a field dominated by men. At the same time, she willingly downplayed her skills and played up her femininity to protect the cause of aviation. And while she did challenge gender barriers, she fought largely those that blocked her own personal success.

During World War II, Cochran's conservative approach led her to conform to gender stereotypes. She wanted a place for women pilots in the war, but was an adamant voice against women in combat. She assured the public that women's roles were limited and temporary. According to Cochran, WASPs would only release men for combat, rather than replace them. She did, however, regret that she could not fly combat missions. "If I were a man," Cochran commented to Chuck Yeager sometime after the war, "I would've been a war ace like you. I'm a damned good pilot. All these generals would be pounding on my door instead of the other way around."[7] Cochran had similar feelings during the space age. In the 1960s, Cochran concluded that although women were not ready to be astronauts, she nonetheless wanted an opportunity to go into space.

* * *

Jacqueline Cochran was one of the best-known women in America during the 1930s and 1940s. Unlike many prominent women, she was neither a feminist activist nor a political reformer.[8] She achieved great fame, but did not embrace the struggle for equality. Cochran is, however, an example of the "New woman" who appeared after World War I. Driven by a desire for independence and personal freedom, she spent the 1920s

Introduction

as a single beautician, living on her own before achieving wealth and fame. Cochran was similar to many women who became famous, whether as movie stars, sports heroines, artists or career women — women such as Amelia Earhart, Babe Didrikson, and Gloria Swanson. According to Susan Ware, these women thrived on publicity because it made their careers possible. Meanwhile, Americans took notice because they had more leisure time and women were interesting.[9] Cochran fits into this category well. She was a glamorous, movie-star-like aviator with an amazing ability to publicize herself. She was frequently mentioned in the *New York Times*, beginning in 1932 when she earned a pilot's license in two-and-a-half weeks. Later, she was in the spotlight as a woman air racer who established aviation records and won many awards.

Historians have paid little attention to Cochran, or to women pilots in general. Much of what has been written about women pilots has focused on Amelia Earhart. Mary S. Lovell's *The Sound of Wings* and Susan Ware's *Still Missing* are the best of these studies, and Ware's book is especially important for its discussion of Earhart's feminism.[10] In 1998, Leslie Haynsworth and David Toomey in *Amelia Earhart's Daughters* discussed women aviators from the World War II era to the Space Age. The first chapter is devoted to Jacqueline Cochran and several chapters discuss the WASPs Program. Cochran is also discussed in a 1993 popular history entitled *Women with Wings: Female Flyers in Fact and Fiction*. In the 1980s, Janet Dailey published a historical romance book which revolved around the WASPs Program. In Dailey's *Silver Wings, Santiago Blue*, Cochran appears as a historical character championing the cause of women in aviation. In 2000, Margo McLoone published a small book entitled *Women Explorers of the Air*. Cochran is one of five women aviators profiled.

Also important for its analysis of women in aviation is Joseph Corn's *Winged Gospel*.[11] Corn explores America's fascination with aviation and the airplane's place in society. He also analyzes gender stereotypes in the aviation industry and how women were used as advertising tools to sell aviation safety.

Studying Cochran's flying career from 1932 to 1945 can help build on Corn's path-breaking work and move the study of women and aviation beyond its focus on Amelia Earhart. Cochran is most important because of her role in organizing women for war. Hence, a study of Cochran also contributes to the literature on women and World War II. Cochran's postwar career adds to the study of women in male-dominated professions. Cochran's unsuccessful 1956 run for the House of Representatives adds to the study of women and minorities in politics. Her records made in jet-propelled aircraft and her involvement in America's space program, chronicles advancement in aviation. While Cochran never became an astronaut, she was involved in testing women's ability to go into space. A study of Cochran adds to the study of women in space exploration.

Only in the last 20 years have historians begun seriously exploring the topic of women in World War II.[12] With the exceptions of Mattie Treadwell's *The Women's Army Corps* and D'Ann Campbell's *Women at War*, most works deal with women in industry.[13] Recently, several works have been published on the Women's Air Service Pilots. Some are memoirs, such as Byrd Howell Granger's *On Final Approach*, Adela Riek Scharr's *The WASP: Sisters in the Sky*, and Marion Stegeman Hodgson's *Winning My Wings: A Woman Airforce Service Pilot in World War II*. Others, like Marianne Verges's *On Silver Wings*, and Jean Hascall Cole's *Women Pilots of World War II*, offer brief, popular accounts of the program.[14] Recently, Molly Merryman published *Clipped Wings*, which concentrates on the program's disbandment.[15] In 2007, Doris L. Rich, the author of several books on women aviators, wrote a biography of Cochran. The book, *Jackie Cochran: Pilot in the Fastest Lane*, highlights Cochran's gutsy, "go-getter" attitude.

Cochran's work with the WASPs is an example of the unrealized and little-studied contribution of women to the war effort. Their story also shows the huge constraints gender imposed on women. Because they were women, WASPs were limited to certain duties and were often distrusted to perform their responsibilities. The WASPs was a civilian, rather than military, organization. After the war ended, the women were not considered veterans and could not receive any military benefits. Cochran accepted and sustained these restraints. She challenged some perceptions of women, yet worked within the system. Without her willingness to accept gender stereotypes, however, women pilots would have had a smaller role in the war effort.

Cochran's contributions deserve greater scrutiny because she helped shape modern aviation and brought greater attention to women flyers. In 1987, Maryann Brinley published a work on Cochran.[16] The book is a collection of oral histories in which Cochran's life is told in a patchwork of snippets. Neither this work, nor Cochran's own 1954 autobiography, place her activities in historical context.[17] As this study demonstrates, Cochran was an important figure in the history of women in aviation and war.

1

The Making of Jacqueline Cochran

Jacqueline Cochran, America's premier woman aviator, never knew her real name. Raised by a family named Pittman, she molded herself into Jacqueline Cochran. Her life was a "rags to riches" story. An orphan, reared in poverty she pushed herself into a better life. By 1938, Cochran was a wealthy woman and an internationally known aviator. Her remarkable story began in the early 1900s and chronicles seventy years of accomplishments by sheer determination.

Cochran came of age in the 1920s, a decade when women were experiencing greater freedom and opportunities. Following World War I, women gained the right to vote, entered professions formerly reserved for men, and achieved fame on the silent screen and in sports. Despite the new opportunities, women's great successes in any field was extremely rare. Cochran's rise was particularly unusual.

The life of Jacqueline Cochran is the story of a woman in search of independence and security. Her relentless drive to better herself was rooted in childhood poverty and a loveless family. Throughout Cochran's adult life she possessed an enormous determination, a daunting capacity for hard work, a huge ego, a keen mind and great beauty. Jacqueline Cochran, however, was the creation of a young girl named Bessie Pittman.

The first 20 years of Jacqueline Cochran's life stand in marked contrast to her later years. Growing up in rural Florida, she worked as a common laborer, beautician, and nurse. By the early 1930s, however, she was a well-paid New York hair stylist who had begun buying her own beauty shops. In 1936, still in her late 20s, she married millionaire businessman Floyd Odlum. Cochran's early life had a decided impact on her future. Her underprivileged childhood shaped her into a courageous, determined woman. It also contributed to a less pleasant side of her personality. Vain, self-centered and brash, Cochran sought to impose her will on others. An ambitious woman, Cochran spent her life determined to escape poverty and the world of Bessie Pittman.

What we know of Cochran's early life comes entirely from her own accounts. Her autobiographical writings are enlightening, albeit self-serving. Cochran's earliest years, even her date of birth, is a mystery. Born sometime between 1906 and 1910, Cochran was raised as Bessie Pittman by her foster parents, Ira and Mary (Grant) Pittman, who moved from one Florida milltown to another. The Pittmans had four children and promised to raise "Bessie" as one of their own: It was not until she was six years old that she overheard Mary Pittman explain that "Bessie" was someone else's child.[1] The revelation stunned her, but having discovered she was an "outsider," she began to take pride in the knowledge.

For the first eight years of her life, Bessie lived in northern Florida in the poor, rural milltowns of Sampson, Bagdad, and Millville. She later described these towns as "a sawmill with appendages," because they all consisted of a mill, a schoolhouse, a doctor's office, a company store and the workers' dwellings.[2] Life in these towns was a dreary struggle. Sampson, for instance, was a town with 200 people, a Baptist church, a public school and four general stores.[3] The Pittmans lived in a drafty shack with paper covering the windows, and few furnishings. The children slept on cold pallets on the floor. When Ira Pittman contracted typhoid fever, his illness threw the family into dire hardship; the entire family had to subsist on a dollar a day, the joint income of the two Pittman boys.[4]

Struggling to survive, the Pittmans failed to develop close family relationships. Bessie described her foster mother as lazy and mean. "The knowledge that I did not belong to her gave me a sense of happiness and exhilaration," Cochran commented years later.[5] There was little, if any, love between them. Once when Bessie won a porcelain doll in a Christmas raffle with a ticket she earned, the Pittmans took the doll and gave it to her older sister's baby. Later events revealed how deeply this act stung Bessie. Some 20 years later, when the sister Willie Mae asked for financial help, Cochran agreed if Willie Mae would return the commis-

1. The Making of Jacqueline Cochran

sary doll, won some twenty years earlier. When Cochran died in August of 1980, her doll was buried with her.[6]

Trapped in poverty with an unloving family, Bessie yearned to escape. When she was sent to the local school at Bagdad, she only stayed three days, and took instead to wandering through the woods and watching the freight trains with their string of boxcars advertising "far-off places." Escape, however, was not easy. She tried to join a touring circus, but it would not have her. Neither would a band of gypsies, who chased her away.[7] Soon, however, Bessie found if she could not leave, she could gain autonomy by earning her own money.

Starting at age seven, Bessie worked almost continuously, rarely attending school. Her first job was as a nursemaid for pregnant milltown women; for this she earned ten cents a day. She cooked, cleaned and cared for the expectant mother, as well as the other children. Once, while Bessie cared for a patient, the young mother went into labor and Bessie delivered the child. The experience helped shape Bessie's future. She was determined not to fall into the same trap of early marriage and motherhood. Even at a young age, Bessie knew that the life a small-town girl was not enough for her.[8]

In her quest to escape poverty and the Pittmans, Bessie learned valuable lessons from a schoolteacher and a Catholic priest. In Bagdad, she was entranced by the new teacher, a Miss Bostwick. Years later, she referred to her as "the most beautifully dressed woman I had ever seen ... [and] the greatest influence in my early life."[9] Miss Bostwick saw in Bessie a little girl who needed attention and financial help. She paid her ten cents a week to bring firewood to her room, where Bessie would spend many hours a day.[10] Miss Bostwick helped her improve her reading and even gave her her first book, Charles Dickens's *David Copperfield*. She struggled through the lengthy text and was determined to read every word. When she came to a word she did not know, she would copy it on paper for further study. The next day she would meet with the teacher who would tell her how to pronounce the word and explain its meaning.[11]

Miss Bostwick also gave Bessie the love and care that she never received from her foster parents. She bought the little girl her first new dress and gave her a comb and hair ribbon. More importantly, she taught

Bessie about personal grooming, how to keep her clothes and body clean and fix her hair. Every day, Bessie carried buckets of cold water to the family washtub and bathed. Her family members made fun and accused her of "putting on airs." The strong-willed young lady ignored the snide attacks and persisted.[12]

About the same time, Bessie began attending a Catholic church. The Pittman family never attended church, but they encouraged Bessie to go to Mass. Her foster family, Cochran later assumed, was fulfilling a promise made to her real mother. She described the Catholic priest who visited once a month as "good man who represented something big and wonderful and wholesome to me and I learned about godliness from him."[13]

When Bessie was about eight years old, the Pittmans left Florida for the cotton mills of northern Georgia. The Pittmans viewed this move with hope because unlike working in the sawmills, the entire family, including Bessie, could work in a cotton mill.[14] Following a miserable train ride, the family settled in Columbus, Georgia, and moved into a company milltown house with running water and a bathroom, the first bathroom Bessie had even seen. They soon purchased furniture on credit, knowing the local cotton mill would provide the entire family with work.[15]

Work in the cotton mills was grueling, especially for women and children.[16] Bessie worked the 12-hour night shift and earned six cents an hour delivering spools of bobbin to the weavers. Shorter than the cart she pushed down the aisles, she stacked the spools of bobbin at the sides of the cart to see where she was going. Despite the long hours, poor lighting, and bad ventilation, Bessie loved working at the Columbus cotton mill. Her supervisor, recognizing her as a valuable worker, promoted her after two months to the job of repairing the warp.[17] Soon thereafter Bessie moved from the mill floor to the inspection room, where she examined the cloth for flaws.[18] Before she left the mill, she was placed in charge of 15 other children as an inspection-room supervisor. She taught the other children how to inspect the cloth and repair the flaws. "Being the boss was a role I reveled in," she once recalled. "It came naturally to me even when children older than I was were in my crew."[19]

Despite the exploitation, the mills did offer some people a better

1. The Making of Jacqueline Cochran

life. Bessie learned the importance of making money. She spent two years of her childhood in the Columbus cotton mill, waiting for a better opportunity. Bessie viewed her mill job, at least in retrospect, not as exploitation but as a stepping stone to better things. Unlike the majority of the other workers, she knew that one day she would escape this life, leaving the others behind.[20] "I became self-supporting," she reflected years later, "and was on my way to independence, and I was supremely happy."[21]

However happy she may have been, her chance for "better things" came because of a disastrous textile strike. Seeking better conditions, the mill workers formed a union and went on strike. The strike, which included twenty-seven mills, lasted for three months. With no money to pay rent, the workers were in danger of eviction. Several riots broke out. Bessie eagerly picketed and threw bricks along with her fellow workers. She did not understand why the strike was necessary; she simply hoped to win back her good job.[22] Impatient and in need of money, Bessie left the mill and went to work for a beauty shop in Columbus, running errands and cleaning up the shop.[23]

Bessie could hardly have known it at the time, but she was beginning a lifelong career in the beauty business. Soon she was mixing shampoos and dyes and giving customers permanent hair waves. As in the mill, she worked hard, learned on her own, and made money.

After working in the shop for two years, Bessie almost lost her job when Georgia passed a child labor law governing children under 14. The law, however, was easily evaded when the shop owner lied about Bessie's age to state investigators.[24] Always eager to gain any advantages, Bessie exploited the lie by threatening to expose it unless she was made a full-time operator at $35 a week.[25] After a year in Columbus, Bessie left for a department store salon in Montgomery, Alabama.[26] She landed the job because, in the early 1920s, permanent waves were fairly new and few beauticians possessed the needed skills. Working on commission, she soon earned as much money as the manager. "I had arrived at a state of independence," she recalled. "I now had a profession and not a trade."[27]

Bessie's stay in Montgomery brought greater personal enrichment and opportunity. She earned money, learned to dance, dressed in stylish clothes, bought a Model T Ford, and dated young college men. Her first customer, a Mrs. Lerton who was a judge in the Juvenile Court, helped shape her life. Bessie had impressed her with her knowledge of literature (most likely gained from Miss Bostwick) as well as her skill of permanent waving. Lerton became a teacher of sorts, directing Bessie's readings, teaching her how to cook and sew and introducing her to new people and new experiences.[28]

Like many young, single women in the 1920s, Bessie sought liberation, but unlike most she did not intend to abandon the idea of working for herself. Having embarked on a lifelong quest for autonomy, she chose not to marry. Instead, Bessie molded herself into the modern career woman, supporting herself and creating her own identify. She did, however, consider another profession open to women — nursing.

At Mrs. Lerton's urging, she abandoned hair styling for the nursing profession.[29] At first, Bessie resisted the idea because she feared her lack of education (she had only reached the third grade) would prevent her from mastering college work. She eventually agreed to enter the program, in part because of her gratitude toward Lerton.[30]

Bessie spent three years learning nursing skills.[31] As she feared, she fell below the other students in academic standing, but she used her near-photographic memory to compensate for poor education. "My memory was perfect," she commented, "and once I had set things up for an operation, I knew exactly what was wanted and when."[32]

In nursing school, Bessie found a sense of belonging. A Catholic institution, the Montgomery Hospital was staffed with nuns and other practicing Catholics. Bessie used this opportunity to strengthen her own faith and connect with her past. She believed that her real parents, whoever they might be, were strong Catholics. While Cochran never knew her heritage and often claimed she did not want to know, she clung to the Catholic religion. It was her only lifelong link to the past.[33]

Bessie enjoyed her years at the Montgomery Hospital. But, unable to master the academic side of nursing school, she never officially graduated and would not take the State Board Examinations. Bessie's fear of failure was greater than any desire for a degree. Besides, she knew a place that needed nurses, with or without degrees — Sawdust Road.[34]

Bessie's childhood reminiscences had revolved around one major theme: escaping "Sawdust Road." Now 13 years after leaving the sawmills, she decided to return. No longer poor, she felt compelled to help the

1. The Making of Jacqueline Cochran

less fortunate and accepted a nursing position as "district nurse" in Bonifay, Florida, a small sawmill town. "I was going back to Sawdust Road again," she commented, "but this time as a modern Florence Nightingale to help the people I knew needed help so much."[35]

Bessie arrived with purpose, yet her hopes were soon stifled. She delivered babies and assisted the doctor in performing operations. Bessie earned three dollars a day, which made her well paid in the milling community.[36] Bessie pitied the overworked doctor, whom she guessed once had "had great ambitions." His clinic overflowed with medical instruments which she found "dirty and rusty and unsterile," a good description of the condition in the doctor's office in general. Appalled by the grimy conditions and overwhelmed by the number of patients, Bessie wondered if nursing was worth the struggle.[37] Frustrated and unhappy, Bessie abandoned "Sawdust Road," never to return. Soon she would rid herself of her last vestige of her unpleasant youth.[38]

Jacqueline Cochran sets new world U.S. Women's Speed Record in 1937 in a Beechcraft (photograph courtesy of Smithsonian Air and Space Museum, SI-77-4171).

* * *

Bessie spent the next few years restlessly searching for a new path. From Bonifay, she moved to Pensacola, where she resumed her beauty career and shortly became a partner in a shop. Determined to break completely from the past, she chose a new name. Bessie Pittman disappeared; Jacqueline Cochran emerged. According to Cochran, the name was chosen randomly from a phone book. However, another source claims that her unwed mother had already given her the name Jacqueline. The desperate teenager left her baby with the Pittmans with a note reading, "She's a good baby. The night she came I dreamed she climbed clear to heaven like Jacob's angels. So I named her Jacqueline."[39] Whether Bessie chose both names or only the name "Cochran," this decision was a significant event in Cochran's life. Shunning the Pittman name, she sought to remake herself and develop a new identity. As Jacqueline Cochran, she would find the freedom, independence, and personal fame she had long desired.

By coming of age in the 1920s, Cochran could take advantage of the new opportunities for women. The decade brought dramatic changes in women's lives, starting with the right to vote.[40] From there, some women became social activists, took up causes, pressed for equal rights and social reform. According to Dorothy Brown, the "new woman" assumed several guises, including veteran reformer, victorious suffragist, pioneering scientist and more.[41] Others, particularly younger women, concentrated on self-fulfillment over social service.[42] With the fight for suffrage over, women began liberating their personal lives. They threw away long dresses and petticoats, bobbed their hair and became self-supporting, some as professionals in areas formerly reserved for men. After World War I, more American women sought financial independence and entered the work force.[43] While most women were confined to "women's professions," opportunities in these areas increased. According to Glenda Riley, "The spirit of the 1920s encouraged women to express themselves as independent, sexual, and political individuals."[44]

This liberated woman of the 1920s took several different forms. She could be the carefree flapper, smoking, drinking, palling around with men, or the self-supporting, independent businesswoman, or the New Style feminist who attacked prewar feminists as unfeminine and asexual.[45] According to writer Dorothy Dunbar Bromley, the new woman was well dressed, enjoyed the company of men and believed that full life calls for marriage and children as well as a career.[46]

Cochran led an independent, free-spirited life. She was a self-supporting woman traveling the country alone, all the while challenging many barriers. Cochran was, however, a champion for herself and *not* for women's rights. "About women," Cochran confided years later, "I don't think too much of 'em. I'd rather spend an evening with a man any time than a woman."[47] There were several reasons for Cochran's lack of enthusiasm for feminism. While most women who were caught up in social changes were from the urban, middle-class, Cochran was a poor country girl with few opportunities. As a young woman growing up in the rural South, she was not exposed to feminist ideas or to other women involved in the feminist movement. Despite grandiose plans for herself, Cochran took a cautious stance when it came to advancing the positions of other women.

The desire for independence was a guiding force in Jacqueline Cochran's life. She, however, was no feminist and abhorred the label. While Cochran enjoyed the new freedoms open for women, she did not crusade for greater equality; nor did she pursue the fads of the 1920s. Cochran did not bob her hair, attend "petting" parties or frequent speakeasies.[48] But she did take advantage of the greater economic opportunities for women. As Cochran had learned, money meant security, and security meant independence.

Cochran's decisions were rooted in the hated experiences of her youth. She was determined to have a better life. In some ways, poverty had a positive effect on Cochran's adulthood. Growing up disadvantaged, she could be sympathetic to others' needs. On the other hand, Cochran's harsh childhood made her ambitious, driven and self-centered. These traits would often alienate people and impeded Cochran's lofty goals.

Cochran's ongoing search for autonomy inspired her to travel from one city to the next. During the late 1920s, she worked for a time as a beautician in Pensacola, but soon grew restless and took a job as a traveling saleswoman, peddling dress patterns.[49] As a saleswoman, she enjoyed the freedom of the road but, concerned about her future, she decided to open a beauty shop in Biloxi, Mississippi, at the Edgewater Hotel. When a hurricane destroyed much of the area, Cochran reluctantly returned to the Pensacola shop.[50]

Despite the setbacks, she had a flourishing business and an active social life. She learned to dance and attended dances at the Pensacola Naval Flying School. "Though flying was the farthest thing from my mind then," Cochran recalled, "I took one step in the direction of my later life."[51] She danced with and dated future captains and admirals.

Although she would not take up flying until 1932, she later reflected, "They talked flying. I listened."⁵²

Pensacola did not hold her for long. She spent nine months at a beauty school in Philadelphia where she worked part of the time as an instructor. The beauty profession was one of the new vocations available to women in the 1920s. It was a respectable profession which catered to women by women. By 1930, over 40,000 beauty parlors were open for business in towns and cities across the nation.⁵³ Meanwhile, the movie and advertising industries presented beautiful, youthful women as examples of how every woman should look. To achieve this look, beauty parlors and cosmetics became necessities.⁵⁴ Cochran excelled in the beauty profession, using her good looks as an advertising tool. After Philadelphia, she moved to New York. "I thought I was ready for the biggest city of all," she later said.⁵⁵

Confident of her abilities, she was determined to start at the top. She first went to the famous New York hairdresser Charles of the Ritz and asked for a job. She found Mr. Charles "egotistical and irritating." In turn, he was amused by Cochran's self-assurance and claims of expertise. When Charles told Cochran she could not work for him unless she cut her long hair, Cochran stormed out. The next morning, he had a change of heart, and called to say he would hire her—even with long hair. Cochran declined and took a job with Antoine at his Saks-Fifth Avenue Salon. There, she gave permanent waves and $40 hair dyes. In the 1930s, beauticians who could give permanent waves without burning the hair were much in demand. Interested in being more than just a stylist, Cochran began investing in several smaller beauty shops, dividing her time between Antoine's and these other businesses.⁵⁶ After working the entire day at Antoine's Cochran would go to one of her own shops, where she was part-owner, and supervise the operators.⁵⁷

Cochran enjoyed working at the salon and began earning a great deal of money. Her new position also exposed her to a very different social set. Despite her harsh upbringing, she considered herself lucky. She always seemed to be in the right place at the right time and had a talent for meeting influential people. While working as a beautician, Cochran met a rich businessman who was in a failed marriage. Floyd Odlum, a reserved, ambitious millionaire, would become central to her

1. The Making of Jacqueline Cochran

life. The two dated secretly for several years and married in 1936. With his financial assistance, Cochran was able to build a cosmetic company, buy expensive airplanes and enter air races. Odlum was an essential part of Cochran's success.

In less than 30 years, Cochran experienced an amazing rise from a dirt-poor girl to a successful Fifth Avenue stylist; someone with money and influential friends. Her unusual good fortune came through at the time when most people were struggling through the Great Depression. Despite her great success as a beautician, she continued to search for freedom and adventure. Three years after moving to New York, Cochran temporarily left the beauty business to take up flying. Shortly after earning a pilot's license, she established a cosmetics company. In the midst of economic turmoil, she was about to embark on two new careers — one was to manufacture beauty projects and run her own salons; and the second to become the best woman flyer of the 1930s.

2

The Lady Is a Pilot

By the 1930s, Jacqueline Cochran's life had taken a dramatic turn. Now far from the Florida milltowns, she established herself as a successful New York hair stylist. Cochran, however, was not content to remain an obscure beautician. Her desire for fame and adventure led her to become an aviator — the best woman aviator in the world.

Cochran earned her pilot's license in 1932. This was an act of independence for Cochran, and it suggests that, even during the depression, some women continued to pursue social liberation. In the 1920s many young women took up sports, drove automobiles, and a few flew airplanes. They had bobbed hair, wore knee-length dresses and donned slacks. Jacqueline Cochran benefited from these changes, particularly those involving women and sports.

Women who played sports directly challenged the idealized version of the Victorian woman. According to Dorothy Brown, "The stereotype of the weak and shrinking violet was challenged by a cadre of athletic champions."[1] Tennis stars Hazel Wightman and Helen Willis and swimming champion Gertrude Ederle proved that women could be athletic, strong, and courageous without losing their womanhood. At the same time that women were swimming the English Channel and winning Olympic gold medals, they also took up the sport of flying. Stunt pilots Ruth Law and Katherine Stinson did wing walking and aerobic loops for air shows. In 1928, Amelia Earhart captured national attention when she flew across the Atlantic. Even though Earhart was a passenger, her flight caused a sensation because she was a woman who resembled aviation hero Charles Lindbergh. Earhart soon became known as "Lady Lindy," the female symbol of a daring aviator. By the time Cochran earned her pilot's license, it was generally accepted that women could fly airplanes — as long as they did it for sport and not as a profession.

Cochran's plunge into aviation was sudden, swift, and enduring. She earned a pilot's license in three weeks' time and, shortly afterwards, joined an air circus and began entering air races. Beginning in 1932, Cochran's life would be forever entwined with aviation. As a woman of action, she used flying to accomplish other lofty goals — a life of freedom, adventure, and fame.

* * *

Jacqueline Cochran became a pilot during an important time in aviation. In the 1930s, industry, the government, and many pilots made a concerted effort to make flying more than a novelty. Air races and shows sought to revive public interest in aviation that had waned after World War I. Meanwhile, industries like Boeing and Douglas introduced commercial airlines.

The airplane industry took hold in the 1930s, although it had been building since the early 1900s. By 1911, approximately 12 airplane manufacturing companies existed and 50 more produced aircraft parts and supplies.[2] Some of the early manufacturing founders included Glenn Curtiss, Alexander Graham Bell and Orville and Wilbur Wright. The airplane industry was briefly stimulated by American's entrance in World War I. The government ordered the production of over 20,000 airplanes for the war effort. By 1919, America had 31 companies producing aircraft and parts.[3] In the immediate postwar period, the aircraft industry suffered from a lack of interest then rebounded because of the growing usefulness of the airplane. According to Roger Bilstien, the airplane "proved useful in dozens of ways: private business and corporate flying, cotton dusting, seeding, photography, and others."[4] In 1921, manufacturer Donald Douglas suggested that the airplane could be used to haul freight. Meanwhile, other manufacturers pondered the possibility of commercial air travel, especially for the busy executive. In 1925, the airline industry began to flourish when the government established contract mail routes with commercial carriers.[5] Two years later, popular interest in aviation reached unprecedented levels when one of the government's appealing air mailmen, Charles Lindbergh, soloed across the Atlantic in his airplane *The*

Spirit of St. Louis. Lindbergh's well-publicized flight brought greater attention to aviation.

In the 1930s, manufacturers made improvements in aviation designs, which resulted in faster, more reliable airplanes. Benefitting from the technological advances of the 1920s, innovations continued in spite of the depressed economy. Ideas developed during the 1920s and became realities in the 1930s. New designs that featured cantilevered wings, retractable gear and stressed-skin construction accommodated innovations such as streamlining, variable-pitch propellers, wings flaps and engine cowlings.[6] Meanwhile, new engines, such as the Wright J-5 Whirlwind 220-horsepower radial air-cooled engine and the 400-horsepower engine and the 400-horsepower Wasp improved speed and carrying capacity.

The 1930s was also a significant decade for passenger airplanes. Commercial airlines introduced new airplanes, including the Boeing 247, DC-3 and DC-1. Meanwhile, airline companies advertised the comfort and conveniences of air travel in an effort to persuade the public to fly. Despite the DC-1's modern conveniences — hot- and cold-running water, upholstered seats and air conditioning — air travel was unpopular with the general public. Most people considered the venture far too dangerous. But, slowly, the public warmed to the idea of flying.

Beginning in 1927 and into the 1930s, a revival of "air mindedness" (the public's fascination with flying) took place. "To be 'air minded,'" suggested Joseph Corn, "meant having enthusiasm for airplanes, believing in their potential to better human life, and supporting aviation development."[7] Following World War I, America had become disenchanted with aviation. The airplane seemed impractical in peacetime and, not surprisingly, automobile construction took precedence over the aviation industry. The government sold surplus airplanes "at bargain prices." Meanwhile, aviators found themselves unemployed and looking for chances to fly.[8] According to one former armed forces pilot, "The greatest hazard in aviation is starving to death."[9]

In the 1920s, few jobs existed for aviators. Some men flew for the Army or the Navy, but at the time military aviation was struggling "to find its place in America's defense establishment."[10] Others started their own businesses, giving lessons or offering airplane rides for three to five dollars. Aviators Wiley Post and Charles Lindbergh became mail pilots. The United States Air-Mail service officially began in 1918. Mail pilots transported personal letters as well as stocks, securities, and commercial correspondence. In 1925, after the passage of the Kelly or Airmail Act, private lines carried mail under post-offices contracts.[11] Finally, as commercial aviation was established, aviators found work as airline pilots.

Pilots, frustrated by the lack of opportunities, looked for ways to revive air mindedness in America. If the public took a greater interest in aviation, the demand for pilots might increase. Furthermore, the public needed to become more "air minded" if aviation was to become a major industry. To help generate more interest, some pilots spoke out on the importance of aviation; others took up barnstorming, air racing and aerial publicity stunts. The pilots themselves stimulated interest. Indeed, the aviators of the 1930s were the media darlings of their day. The people loved them, the press glorified them and the aviation industry benefitted.

Several pilots enjoyed popular appeal. They included the one-eyed Wiley Post, who flew around the world in 1931, and the flamboyant Roscoe Turner, whose flying companion, a lion named Gilmore, generated as much excitement as Turner's considerable flying skills. Turner even fashioned Gilmore with a special parachute in case of emergencies. America remained fascinated with Billy Mitchell, the air hero of St. Mihiel, and was intrigued by the mysterious and wealthy Howard Hughes.

In the 1920s, some military and industrial leaders had made a concerted effort on behalf of aviation. Perhaps the best-known spokesman was the World War I hero Colonel Billy Mitchell. His experience in the war had a profound impact on his beliefs. He had "seen the future and it was called air power."[12] Mitchell and his followers set out to prove the usefulness of aviation in peace as well as war.[13] Mitchell, however, was unsuccessful. It took Charles Lindbergh to bring aviation into fashion. Lindbergh, with his handsome looks, polite demeanor, spotless reputation, and boyish blue eyes, caused a sensation with his 1927 nonstop flight from New York to Paris. This feat did more to stimulate a rebirth of interest in aviation than any other event. While others had crossed the Atlantic before him, Lindbergh crossed it alone. Most scholars agree that this fact was "the essential key to his impact."[14] Lindbergh's flight inspired dozens of aviation songs and books. As one historian noted, "In films, comic strips, and magazine advertisements, aviation emerged as an accepted token of popular culture."[15]

To help revive public interest in aviation, some of these daring pilots took up barnstorming and participated in air races. Barnstorming was a dangerous activity where flyers performed aerial acrobatics and other death-defying stunts, like sky writing and wing walking. The stunts, did, however, serve broader purposes. Aerial acrobatics improved pilots' skills and helped stimulate interest in aviation.[16] These "air circuses" gained popularity after World War I when flying jobs were difficult to find. In their government surplus Jennys, former military pilots and other flyers thrilled audiences.

Air races also helped to stimulate interest in aviation. National Air Races started in the 1920s and included the Thompson Trophy and the transcontinental Bendix. In the beginning, the races were dominated by military pilots, who won most of the prizes. Later winners included civilian pilots and both men and women. Air races attracted huge crowds and broad press coverage. According to Bilstein, "Thousands more followed the races in papers and national journals that also gave daily coverage during the progress of sundry long-distance flights."[17] Air races also encouraged technological progress. They served "as the proving ground for aerodynamics, airplane frame strength, engine endurance, and so forth."[18] In the 1920s and 1930s, manufacturers were designing airplanes that needed to be tested. In the 1930s, aircraft companies often sponsored pilots to fly their airplanes in races.

While most aviators were men, several women also flew. Beginning in the 1920s women earned pilot's licenses, an expression of their search for independence and liberation. While women pilots were often discriminated against, their desire to fly, at least for a few, outweighed the obstacles.

Despite their abilities, women pilots found it nearly impossible to establish careers in aviation. They were excluded from the military and many flying-related jobs. Helen Richey was the only women pilot to break into male-dominated positions. In the mid-1930s, she was the only female air mail pilot and airline co-pilot. In the 1940s, Richey was the first woman to earn an instructor's license.[19]

2. The Lady Is a Pilot

It was a fact of life that women who tried to fly faced hostility and suspicion from men who dominated the new world of aviation. The case of Blanche Scott offers but one example. In 1910, aviator Glenn Curtiss reluctantly agreed to give her some basic instruction, but with the aid of the throttle block, he was determined to keep her grounded. Despite this handicap, Scott took off and climbed 40 feet into the air. Eventually Curtiss relented and even asked her to join his exhibition team. Scott became an outstanding stunt pilot, working the air shows for six years. Nevertheless, she became disillusioned with flying, quitting at age 27. The public, she felt, failed to take women pilots seriously. "Too often," Scott sadly confessed, "people paid money to see me risk my neck more as a freak — a woman freak pilot — than as a skilled flyer."[20]

To dispel any doubts about their flying abilities, women pilots set out to make new records and match "the men's fate-tempting feats."[21] In 1912, Harriet Quimby, the first American woman to earn a pilot's license, became the first woman to fly across the English Channel. Three months later, she was killed when turbulence threw her out of her airplane. Her tragic death may have added to the idea that flying was too dangerous for women.

Despite the limitations put on women pilots, Quimby saw a future for women in aviation. Shortly after her death, *Good Housekeeping* magazine published an article in which she had encouraged women to fly. "In my opinion," she wrote, "there is no reason why the aeroplane should not open up a fruitful occupation for women. I see no reason why they cannot realize handsome incomes by carrying passengers between adjacent towns, why they cannot derive incomes from parcel delivery, from taking photographs from above, or from conducting schools of flying. Any of these things it is now possible to do."[22]

In America, the first woman aviator to fly was Bessica Raiche, who made her first flight in 1910 with her husband. They started the French-America Airplane Company and designed and constructed airplanes with the Wright Brothers, among others. Eventually she gave up aviation for medicine and became a doctor, but not before establishing the trend for donning riding breeches in place of long skirts when flying.[23]

In the years leading up to World War II, several women acted on their strong desire to fly. In 1917, stunt pilot Ruth Law became the first woman to fly at night and the first women to loop the loop. Fellow stunt pilot Katherine Stinson gained fame by designing her own stunt — the dippy twist loop — and flying 610 miles to break the nonstop distance record.[24] When World War I broke out, Stinson and Law volunteered to become military pilots. Despite their considerable skill and experience, the Army promptly turned them down.[25] While many women assumed men's jobs during the war, flying for the Army Air Corps was out of the question.

In the late 1920s and early 1930s, a new generation of women earned their pilot's licenses and searched for a place in aviation. They came from various backgrounds; some were small-town girls and others were well-educated society ladies. Most were married. Some women pilots had spouses in aviation, like Smith graduate Anne Morrow Lindbergh, and Vassar graduate Nancy Harkness, who married pilot Robert Love. Ruth Nichols, a graduate from Wellesley, began flying in 1920. Two years later, she served as a co-pilot for the first nonstop flight from New York to Miami. Throughout the decade, she set several aviation records, flew over 140 different aircraft models and became the first woman (and only the third person) to land an airplane in all 48 states.[26] In 1924, Arkansas native Louise Thaden earned her pilot's license and became a manager of an airplane sales company. In 1929, she set a new altitude record for women and also established speed and endurance records.[27] Other women pilots included Laura Ingalls, a former vaudeville dancer, who set records in distance and endurance flights and stunt flying, and Helen Richey, the first woman mail pilot and airline co-pilot.[28] Richey dreamed of being either a stunt or test pilot, but no aircraft companies would hire her. The most visible woman flyer was Amelia Earhart, whose uncanny resemblance to Charles Lindbergh and media-hound husband, publisher George Putnam, helped make her famous.

By the time Jacqueline Cochran earned her pilot's license, women in airplanes were not unusual. These women pilots reflected changing roles in gender as America moved from the Victorian era to the liberating 1920s. Seeking opportunities beyond housekeeping, women, especially educated, young, middle-class women, sought a means of freedom. Flying was the ultimate symbol of freedom and liberation in the 1920s. According to Joseph Corn, "It was this giddy sense of liberation they found in the sky which prompted so many women to predict that the new field of aviation promised great opportunities to their sex in the future."[29]

While these women possessed a great desire to fly, their progress was hindered by gender discrimination. The airplane was viewed as a dangerous machine that women were physically and emotionally unable to handle. According to Susan Ware, the aviation "industry developed along sex-segregated lines that marginalized women. They were welcome as stewardesses but banned as pilots; they could demonstrate light sports craft but were denied access to heavier commercial and military aircraft; except for an occasional woman in the front office to deal with the 'women's angle' they were frozen out of the business side."[30] When air mindedness finally came back into fashion, women were not thought of as pilots but as advertising tools used to convince others to fly.

When commercial aviation emerged, many people were afraid to fly. The public's perception of aviation was that of an activity reserved for certain people. Enthralled with the daring feats of World War I "aces" and captivated by the postwar daredevils, the public believed only a special type of person could fly an airplane. After all, anyone who could soar like the birds must be endowed with some special skills, as well as a willingness to take dangerous risks. Flying as a passenger seemed equally hazardous. The aviation industry was faced with the crucial dilemma of how to ease people's fears about flying. Without a confident public willing to use air travel, the commercial aviation industry could never succeed. One way of reassuring the public was to use women to sell and publicize aviation.

The aviation industry used gender stereotypes to its advantage. Exploiting the perceived weakness of women, the industry encouraged women to promote aviation, and created new opportunities for them. Demonstrating and selling airplanes became a common job for women pilots. Many famous female aviators worked in the sales market, including Earhart, Nichols and Thaden.[31] Even as a mail pilot, Helen Richey found herself performing public-relations duties for her employers. Unlike the other pilots, Helen was expected to attend luncheons and speak at local schools. The visibility of women pilots made flying appear safe. If frail, hysterical creatures like women could fly airplanes, so the argument went, anyone could.

While women pilots resented the stereotyping, they chose not to challenge the industry. Some women, like pilot Louise Thaden, reluctantly agreed with their reasoning. "If a woman can handle it [an airplane]," she complained, "the public thinks it must be duck soup for men."[32] To benefit aviation, women pilots allowed the industry to use them. Jacqueline Cochran also became a willing participate in this game, with little regret. She cared more for aviation than she did for the advancement of women.

Women pilots found themselves in a difficult position, torn between bad publicity for themselves or bad publicity for aviation. According to Joseph Corn, "They often chafed at restrictions on what they might do as pilots. But complaint or protest was difficult, for they also considered themselves evangelists of aviation and sought to do nothing that would in any way damage the cause."[33] In the end, women pilots decided that aviation took precedence over equality. They often downplayed their abilities to further the cause of aviation.[34]

One of the chief spokespeople for women in aviation was Amelia Earhart. She encouraged women to fly as passengers, hoping many would become pilots and bring recognition to female flyers.[35] In 1929, women flyers were confronted with the question "Does aviation offer a career for women?"[36] While most of them agreed that flying for women should be more of a sport than a profession, Earhart believed a woman with the right training and opportunities could excel in aviation.[37] "Exceptional ability always reaches the top," commented Earhart.[38]

Earhart was a liberal feminist, who believed that her individual accomplishments represented a victory for all women. While her aviation feats may not have inspired all women to take up flying, Earhart presented herself as an example for other women to follow. According to biographer Susan Ware, Earhart believed that "women can achieve whatever they set out to do and ... it should be the ability of the individual, not the sex that counts."[39]

Despite Earhart's avowed feminism, she allowed herself to be used as an advertising tool for commercial aviation. She endorsed several airlines, making goodwill tours to promote their safety.[40] With Earhart as spokeswoman, the industry had the voice of a famous flyer as well as a woman. "My job was to sell flying to women," Earhart commented, "both by talking about it and by watching details of handling passengers, which were calculated to appeal to feminine travelers."[41] Earhart did, however, combat gender stereotypes about women by stressing that females could fly as well as male pilots.

Unlike Earhart, most women pilots were not feminists, but their desire to fly was rooted in a need for liberation and their independence to break free of Victorian restraints. These women pilots envisioned a life for themselves beyond the home. They desired adventure, personal accomplishment, even fame. Amelia Earhart later titled her autobiography "For the Fun of It," explaining why she took up flying. In the air, women felt free from society's constraints. And, not unlike men, they wanted to master this machine — the airplane. The best example of this type of woman pilot was Jacqueline Cochran.

* * *

Jacqueline Cochran's flying abilities were unmatched by other women. For Cochran flying was more than a sport, it was a form of self-expression. "Flying offered me the most stupendous sensation," she commented years later.[42] Cochran found in flying a career, an escape from boredom, and a pathway to fame. From the beginning, she viewed aviation as a career which suited her personality. Cochran believed that because she was mechanically inclined, had strong physical endurance and was fearless, she would make a good pilot. "I decided there was a real place for women like myself in aviation and I'd take it up as a profession."[43] Flying allowed Cochran to satisfy her restless and independent nature. Her entire life was spent in search of freedom, and what could be more liberating than becoming an ace flyer?

Cochran's rise in aviation was an unusual accomplishment. She earned her pilot's license in three weeks' time and, within a few years, could outfly other women pilots and most men. How did Cochran become an ace flyer in such a short time? Her driving ambition was an important element in her success. Cochran always pushed herself to be on top, never willing to be second best.

Cochran's willingness to take risks was a second factor in her success. Fearless and daring, she performed dangerous stunts, flew in damaged and experimental airplanes, and even flew in bad weather. Cochran had a flare for danger. When Cochran became an aviator, aviation was in a transitional phase. According to journalist Sherryl Connelly, "The planes of the day, of course, were unpressurized and navigation was a little known art. It was an era in which, in terms of personal safety, you had to be fool to fly and Cochran quickly became a flying fool."[44] Shortly after earning her pilot's license, Cochran barnstormed with an air circus and then entered dangerous air races. According to Cochran, these risks were necessary to improve her skills.[45] Still, she obviously enjoyed the challenge.

Cochran's life was decidedly different from that of most women in the 1930s. While many women were struggling to "make do," Jacqueline Cochran's life was unaffected by the depression. In a time when people refocused their attention on society rather than on the individual, Cochran concentrated on her own interests.[46] She quit her job and took up an expensive sport—aviation.

How did Cochran manage to accomplish such feats during the depression? Cochran's association with Floyd Odlum was a major factor. They met in 1932 while Odlum was separated from his wife, and the two courted for four years. "I kept my confusion to myself," Cochran commented about Odlum's marital status, "and let him do the thinking, the planning."[47]

Odlum was an important influence in Cochran's life. At his suggestion, she turned her thoughts to earning a pilot's license. Odlum believed she could expand her cosmetics company nationwide if she took up flying. Cochran had considered flying before, but had never seriously pursued the idea.[48] Odlum's encouragement, coupled with Cochran's desire to establish a cosmetics business, prompted her decision to learn to fly.

Cochran's decision to take up aviation was a bold idea. She did not, however, intend to challenge the place of male flyers by becoming a mail or airline pilot. Instead, she planned to use flying as a selling and advertising tool. "I never really wanted to copy men or do what men do or should do better," Cochran commented years later. "I only wanted to be myself. And for me that meant flying."[49]

* * *

The idea of taking to the skies appealed to her restless spirit. But she also feared, once again, that her lack of education would make earning a pilot's license impossible. She approached her friend Mike Rosen and asked for his help. Rosen and Cochran had met in New York, where both were beginning their careers. Rosen didn't think she could pass the examination because of her limited schooling. Despite this skepticism, he agreed to help her study for the written examination.[50]

In the summer of 1932, Cochran decided to use her vacation time to learn to fly. She chose the Roosevelt Flying School, located on Long Island. The school offered a free 30-minute lesson and charged $495 for 20 hours of flying time. On the first day, Cochran's instructor, Husky Lewelleyn, informed her that she would have to fly for two or three months before getting a license. When Cochran told him she wanted to finish in three weeks, he laughed.[51]

Lewelleyn and Cochran climbed into a Fleet trainer with a 60-horsepower engine. While confident of her abilities, Cochran winced when she compared the size of the airplane with her "husky" instructor. She was also puzzled by Lewelleyn's teaching strategies. Without explaining the controls or the instruments, he started the engine and took off. Cochran felt exuberant. "I catch the feeling of the plane right away," she commented in later years. "I had just scratched the surface, but I was less beautician and more flyer already."[52] Before the lesson ended, Lewellyen offered a few instructions and let Cochran land the plane. Later that afternoon Cochran began accumulating flying hours. The next day, Cochran's nerves were tested when Lewelleyn demonstrated loops, spins and rolls in an attempt to make her sick. To prove her mettle, she ate a hot dog immediately after he landed the plane. On the third day, Cochran took her first solo flight. Less than three weeks later, with only 18 hours in the air, she earned her pilot's license.[53] Apparently, Cochran had an ability for self-promotion. Her feat made it into the New York papers.

When Cochran arrived at Roosevelt Field on her second day of lessons, she found herself alone. It was seven A.M. and the school did not open until nine. Cochran used those two hours to reflect on her decision to fly. She studied a mural which had been painted by woman pilot Aline Rhonie. The painting showed all the "greats" of aviation" and perhaps Cochran envisioned her own picture among the others. "I wanted to be the world's greatest pilot," she reflected some 40 years later.[54] Within a short time Cochran became an internationally known aviator. Even the skeptical Husky Lewellyen referred to Cochran as a "born flier."[55]

After earning her pilot's license, Cochran foolishly overestimated her abilities. Cochran's first long-distance flight is a good example of her arrogant and impatient nature. Shortly after leaving Antoine's salon, she made plans to attend an air meet in Canada. At this time she lacked knowledge, experience, and an airplane. To plan a route, she bought air maps which she could not fathom. After having the maps explained to her, she consulted more experienced pilots. One "old-timer" advised Cochran to follow the Hudson River to Lake Champlain, and Champlain to Burlington, Vermont, where she should stop and ask directions to Montreal. With little but determination, she set out on her first cross-country flight.[56]

Cochran's flight was made possible by M.E. Grevenberg, whom she referred to as "a gypsy of the air."[57] Grevenberg rented Cochran his Fairchild airplane for $2,000. Cochran failed to mention in her memoir how she had acquired this large sum, since she no longer had a job. Possibly her savings allowed her to rent the Fairchild, but more likely she was financed by Floyd Odlum.

Cochran's flight to Montreal encountered several problems. Fighting poor visibility, she mistook a big canal for the Hudson River and temporarily lost her way. When she finally reached Burlington, Vermont, the airport officials could not believe she needed directions to Montreal, but eventually they supplied a compass, course, and distance. While grateful for the assistance, Cochran did not know how to read a compass. To teach her, several men pushed the airplane around, demonstrating the readings on the compass. Despite these problems, Cochran arrived at the air meet and met up with a relieved M.E. Grevenberg. After the meet, Cochran agreed to fly him back to New York. The return trip was a harrowing experience. "A wall of haze" forced Cochran to land. For two days, Cochran and Grevenberg sat grounded at an unfamiliar airport, prisoners of unfavorable weather.[58] It was an example of Cochran's naiveté that she was dismayed to learn that flying depended on the weather. After Montreal and at Grevenberg's suggestion, Cochran decided to learn to fly "blind"—or by using instruments.[59]

To improve her flying skills, Cochran moved to San Diego, California, and attended the Ryan Flying School. Disgusted by the lack of airplanes and air time, she soon quit.[60] In San Diego, however, she met Navy flyers, one of whom was her old friend Ted Marshall, who offered to teach Cochran to fly the "Navy way." At this time, Navy flyers were some of the best in the nation, and after buying a used Travelier for $1,200 Cochran took instruction from Marshall and seven other Navy pilots for six months. With the help of the Navy and a private math tutor, Cochran earned a commercial pilot's license.

Despite her Navy training, Cochran had still not mastered blind flying. Her desire to do so was only strengthened when she became caught in bad weather while flying Floyd Odlum from New York to Florida. Unable to make turns solely by instruments, Cochran flew straight ahead for 40 minutes. Although she had made a safe landing, the experience frightened her. "I made up my mind," Cochran reflected, "after that experience, that I would never again take up a passenger until I have mastered instrument flying."[61]

She hoped Wiley Post, the flamboyant pilot who had flown around the world in under nine days, would teach her, but he refused. Why would a famous pilot waste his time with an amateur woman flyer? Cochran then tried to teach herself by draping a dark hood over her cockpit and hiring a student to watch for other airplanes while she practiced flying solely on instruments. Eventually, Cochran hired Wesley Smith, a respected transport mail flyer, and the two spent four months and some 500 hours of flying time together. Smith taught her to fly guided by beams, or "sound[s] sent over a certain course through the air by a radio station." Cochran knew she had mastered blind flying when she flew nonstop from Albuquerque to Chicago guided solely on the beam and by instruments. Feeling accomplished, Cochran turned her attention to the MacRobertson Air Race.[62]

Cochran's interest in air races began almost as soon as she earned her license. While she found flying satisfying, she needed the challenge of competition. Prior to the MacRobertson, Cochran had entered a Women's pylon race a Roosevelt Filed in 1933. Although Cochran "placed," she described the Women's pylon as "nothing worth mentioning, either from the standpoint of speed, experience or thrill." Cochran was eliminated in the women's all-around competition because of a poor showing in acrobatic flying.[63] Whatever the disappointments she may have felt from this experience, Cochran made plans for the MacRobertson.

Although she had doubts about winning, she was drawn to the MacRobertson by its $75,000 prize money and her sense of adventure. Racing offered an opportunity to see the world.[64] Air racing also brought publicity, and women flyers were especially interesting to the press. Soon, Jacqueline Cochran would be an aviation headliner.

* * *

Between 1932 and 1934, Cochran transformed herself from a New York stylist to a skillful aviator. Desiring a business of her own, she planned to combine selling cosmetics with flying. In the meantime, Cochran worked as a barnstormer with the Johnny Livingston Air Circus and began entering air races. As an aviator, Cochran found a sense of freedom and liberation. Free-spirited and driven to succeed, she was well suited for flying. By the end of the 1930s, Cochran would become America's top woman pilot.

3

America's Number One Woman Flier

In 1941, the *New York Times* referred to Jacqueline Cochran as the "Number One Woman Flier."[1] In less than ten years she had transformed herself from a successful, yet obscure, New York hair stylist to the most famous woman pilot in the world. Cochran's unusual rise in aviation began in 1932 when she earned a pilot's license in three weeks, a fact noted by the *New York Times*. From there, she joined an air circus and began entering air races. Cochran was not content just to fly. She wanted to be the best, a lofty goal, one that would be duly accomplished in six years.

Between 1932 and 1939, Cochran entered air races such as the 1934 transcontinental MacRobertson, and the annual Bendix Cross Country race which she won in 1938. She successfully flew the dangerous Gee Bee racer (that few men could handle), established speed and altitude records, and risked her life as a test pilot. Cochran became the first woman to make a blind landing, fly military aircraft, and serve on the Collier Trophy Committee. She won numerous awards and was recognized as America's outstanding woman flier for several consecutive years. In short, Cochran surpassed all others — she was America's top female flying ace.

Cochran was one of few women to win an important place in pre–World War II American aviation. Although she faced many barriers because of her gender, Cochran refused to admit this fact. Years later, reflecting on her career, Cochran said that she had never felt discriminated against.[2] In reality, however, Cochran encountered a great deal of discrimination, but she only took notice when discrimination and unfair treatment blocked her career.

* * *

While Cochran may have been a "born flier," it took several years for her to become a skilled one. In 1932, when she made her first long-distance flight, she could not even read a compass. Two years later, she wrecked an airplane in a botched landing.[3] Yet she became a very good pilot in a very short time. In 1934, in fact, Cochran was actually a more skilled pilot than the well-known Amelia Earhart. How did she become such a good aviator? Unlike many women pilots (an exception being Earhart), she had access to the latest airplanes and equipment, partly because of her association with the wealthy Floyd Odlum. Secondly, her willingness to take risks resulted in better skills, as exemplified by her stint with the Johnny Livingston Air Circus. Cochran performed in air shows with the Buffalo, New York "circus," wowing audiences with acrobatic flying. Finally, Cochran's drive and ambition, along with her intense desire for new challenges, were powerful reasons for her success.

Driven by her quest for competition and desire for fame, Cochran flew countless hours and trained under the best teachers she could find. She also gained some experience by flying transport and passenger airplanes. Shortly after earning her license, she secretly flew with the airlines without the passengers' knowledge. She made deals with the airline copilots in which she would take over their distasteful duties, such as serving and comforting passengers, in exchange for a few hours of flying time. As she later explained, "I put in three hundred hours at the airline controls by this strategy and it was the finest sort of training."[4]

Cochran was soon near-perfect in her flying abilities. Her series of aviation records suggest her skill as a flyer. Unlike Amelia Earhart, who enjoyed distance flying, Cochran made a name for herself as a speed racer. In 1937, she set the world's record for women over a three-kilometer course, establishing a speed record between New York and Miami and breaking the 100-kilometer women's national speed record.[5] Cochran continued to set speed and altitude records over the ensuing three decades. In 1939 she became the first woman pilot to make a "blind landing," an accomplishment noted by First Lady Eleanor Roosevelt.[6] The same year, Cochran established a women's altitude record by climbing to 33,000 feet, and broke the international open-class speed record for men and women.[7] In the 1940s, she broke the 2,000-kilometer international speed record and the 100-kilometer national record. Cochran also received the Minneapolis Aquatennial Air Classic Award for being the outstanding woman pilot of 1940, and four trophies from the Women's National Aeronautical Association. In the 1950s, Cochran became the first woman to exceed the sound barrier and the only woman to receive the Gold Medal from the Federation *Aeronautique Internationale*. She also received the French Air Medal, the Lady Drummon-Hay Trophy and Wings of Air medals from Belgium, Spain, Thailand, Turkey and Rumania. In the 1960s, Cochran, enthralled by the jet airplane, set astonishing records. In 1962, she established 69 intercity and straight-line distance records and nine international speed, distance, and altitude records in a Northrop T-38 military jet aircraft. That same year, Cochran became the first woman to fly a jet airplane across the Atlantic.[8] She was also one of a few women to participate in flight testing. In the early 1970s, Cochran was one of three women recognized by the Society of Experimental Test Pilots.[9]

Cochran's work as a test pilot began in the second half of the 1930s and involved high-altitude research and flying high-speed military pursuit planes. Cochran believed the future of aviation depended on advancements in pressurization and the use of oxygen in high-altitude flights. Having flown transport and passenger airliners she had seen many passengers and crew members become groggy or unconscious when planes flew above 15,000 feet. "Once," explained Cochran, "I took a copilot's seat when he passed out from lack of oxygen on a flight between Los Angeles and Albuquerque."[10] Cochran had also experienced the perils of high-altitude flying when she participated in air races. In 1934, during the MacRobertson Air Race, she climbed some 14,000 feet and encountered the same problems faced by all aviators: freezing temperatures and lack of air. Prior to the invention of oxygen masks, pilots survived high-altitude flight by sucking air through a pipe attached to an oxygen tank. At times, this method of breathing was inadequate, as Cochran had discovered in 1937. She climbed to 35,000 feet "in a fabric-covered plane without heating, without pressurization and without an oxygen mask."[11] Because of the lack of oxygen, Cochran became disoriented and ruptured a blood vessel in a sinus. According to Cochran, "All this was part of the cumulative evidence that led up to cabin pressurization and mandatory use of the oxygen mask above certain altitudes."[12]

Since altitude research was in its infancy, few people were conducting experiments in pressurization and oxygen. Cochran finally located a young Mayo Clinic surgeon, Dr. Randolph Lovelace, whose hobby was aviation medicine. After meeting him, she believed that Lovelace and his associates, Dr. Walter Boothby and Capt. Harry Armstrong, held the key to high-altitude flight.[13] For Cochran, their research held such importance for the future of aviation that she eagerly became their guinea pig. When Cochran toured Lovelace's laboratory, she was astounded by his progress in pressurization and the uses of oxygen. As she explained, "Lovelace and this Boothby had actually developed an oxygen mask, a portable oxygen tank, and a low-pressure tank to test individual responses."[14] Cochran eagerly agreed to help test their equipment, including the low-pressure tank and their centrifuge, an apparatus that stimulated gravity with centrifugal force.[15] Later, she conducted test flights for the researchers. "I got involved immediately," she recalled, "for what I could learn for myself as well as what I could offer these doctors."[16]

To understand pressurization, researchers needed to know the effects of high altitudes on human beings and live animals. Cochran aided the research by taking chickens, mice and other small animals up in her airplane and recording their behavior. Once the plane reached certain heights, the chickens would literally burst, leaving Cochran splattered with blood and feathers. "What a mess," she recalled. "The poor buggers would just explode."[17] Sometimes, Cochran reached altitudes as high as 25,000 feet.

By participating in aviation research and setting records, Cochran benefitted aviation as well as her career. As a speed racer, she tested the capabilities of both propeller and jet aircraft. In doing so, she also sharpened her skills and received the fame and recognition she craved. Her records were spotlighted in newspapers and later earned her a place in aviation halls of fame.

Cochran became the first woman to fly military aircraft when, in the summer of 1937, airplane manufacturer Alexander de Seversky asked her to test one of his airplanes. Cochran wanted to fly it in the 1937 Bendix, but Seversky had already promised it to Frank Fuller. Disappointed, she resigned herself to being Seversky's test pilot for the P-35.[18]

In an effort to prove the effectiveness of his pursuit plane, Seversky financed and flew long-distance test flights. Much to Seversky's frustration, Air Corps officials dismissed the tests as "exhibitionist record-breaking" and even called Seversky "undignified" for flying and breaking records as a company president.[19] "To overcome all such absurd objections," explained Seversky, "I arranged to have the pursuit plane flown by a comparative newcomer, Jacqueline Cochran."[20]

Seversky used Cochran's gender to his advantage. He knew that if a woman could fly his P-35, the Army would have a more difficult time challenging its safety. Cochran realized she was being used, but sympathized with Seversky's tactics. She was keenly aware of the skepticism about women pilots and used her flying skills to improve their overall image.

Two days after the 1937 Bendix Air Race, Cochran and Seversky flew to the Wright Field in Ohio. Cochran was elated to be chosen to test this extraordinary yet unpredictable airplane. She also realized that her reputation, as well as the reputation of all women pilots, was on the line: "Sasha knew I was successful flying hot, tricky planes and I think he got the idea that if a woman could fly his P-35 successfully and very publicly it might show up the men with complaints and help him settle his problems with the army brass."[21] But if Cochran "cracked-up" the plane, she knew "they'd blame it on my being a woman."[22] The night before the test flight, an Army officer warned Cochran not to fly the Seversky. "You have no business flying this P-35," he told her. "It's a killer. We have cracked up more than twenty of them."[23] Cochran paid his advice no heed. "Every plane can be flown," she told the brusque general. "And if you can fly a Gee Bee, as I have, you can fly anything."[24] The next morning, with an audience full of Army officers, she climbed aboard the Seversky and took off. The initial flight was successful, but a minor malfunction made landing difficult. Cochran brought the airplane in without knowing if the landing gear was up or down and locked. Fortunately, the wheels were down and she landed without "cracking up" the airplane. Cochran fell in love with the Seversky P-35. "I wanted it desperately," she commented. "I knew it was the fastest thing I'd ever been in."[25] Not only was Cochran the first woman to fly a military aircraft, she was one of the few pilots who could fly the airplane successfully. A few weeks later, Cochran made a speed run in the same P-35, pushing the airplane up to 300 mph.[26] Cochran's performance impressed Seversky but not the Army Air Corps, which continued to believe that the airplane was too dangerous and stopped ordering it.[27]

Cochran also gained fame and experience by participating in widely publicized air races of the 1930s. Air races generated public interest in aviation, gave pilots a chance to prove their skills, and enabled manufacturers to test the durability and endurance of their planes. The first international air meet had been held in Rheims, France, in 1909, with a quarter of a million spectators crowding into the grandstands.[28] In the 1920s, the two major races were the Schneider and the Pulitzer, both of which were dominated by present and former armed forces pilots, and exclusively by men. The first women's air race, the Women's Air Derby, was held in 1929, 20 years after male pilots began racing.

The women who raced were often spotlighted because they were unusual. Their presence, however, brought excitement and interest to the races. In the MacRobertson, Cochran's first race, she attracted attention because she was the only American woman entered. The only other woman entry was the British pilot Amy Mollison, who flew with her husband.

The MacRobertson International Air Race, the world's first classic international air race, attracted pilots from around the globe, including three from America, nine from Great Britain, two from the Netherlands, Australia and New Zealand, and one from New Guinea and Denmark.[29] Named for its sponsor, chocolate manufacturer Sir MacPherson Robertson, the race offered a first prize of 10,000 pounds and a gold cup, valued at 500 pounds. The 16-day event started in England and ended in Australia, covering 11,333 miles.[30]

The MacRobertson included two races, running concurrently. In the speed race, which offered the biggest prize, pilots were judged on "the elapsed time between London and Melbourne." The rules also required pilots to make stops in Baghdad, Allahabad, Singapore, Darwin and Charleville, Australia. In the Handicap Race, which offered a first prize of £2,000. Pilots were judged on actual flying time. "The handicap is figured under a formula which considers rated capacity of engine, weight of plane, wing loading, payload, etc." Jacqueline Cochran, with her crew of Wesley Smith, the crusty mail pilot who had taught her blind flying, and Royal Leonard, who later became Chiang Kai-shek's private pilot, entered both races.³¹

Cochran's original plane was a Northrop Gamma, rebuilt to ensure more power. "[I] installed gas tanks in the mail space," she explained, "to give [it] a 3000-mile range, installed a liquid-cooled Curtiss-Wright Conqueror motor and a specially built supercharger designed to increase the plane's speed at 14,000 feet from 200 miles to 240 miles an hour."³² This quest for speed, however, created serious mechanical problems and Cochran had no choice but to abandon the plane and seek another. She bought a Gee Bee Racer, an airplane known for speed and bad luck.³³ These snub-nosed "bug like" airplanes had a history of instability and were dubbed "widow makers" or "deathtraps." "I think I was one of the few owners of a 'Gee Bee' who didn't lose his life in it," Jacqueline Cochran later commented. "They were squatty, fast, and most unstable."³⁴ "The nearly cute nickname [was] a sham," declared Cochran. "[The Gee Bees] were killers."³⁵ This particular Gee Bee racer was even more dangerous than most.³⁶ The mechanics working on it could not make all the changes before the race began, but Cochran and Smith were determined to enter.³⁷ After a shaky take-off, they settled down for a bumpy flight. "What a ride that was," remembered Cochran, "in an untested and really unfinished plane."³⁸ They successfully soared over England and Western Europe, but over Bucharest their luck ran out. The plane's engine began to splutter, and Cochran made a forced landing.³⁹

Jacqueline Cochran referred to the MacRobertson as "a tragedy of errors" and blamed herself for her poor showing.⁴⁰ She regretted adding the supercharger to her Northrop Gamma, believing the airplane would have flown just as well in its original state. But the experience revealed her determination not just to fly but to achieve a special place as a pilot. Intensely competitive, she fearlessly (maybe recklessly) flew an unfinished plane halfway around the world rather than concede defeat. Once out of the contest, however, Cochran generously aided fellow pilots. For example, she gave Amy Mollison her refueling equipment.

The MacRobertson experience only whetted Jacqueline Cochran's desire for more dangerous racing, and she continued to enter air races until the beginning of World War II. Between 1935 and 1939 she entered the annual Bendix Race, a dangerous, transcontinental speed dash organized by aviation enthusiast and airshow promoter Clifford W. Henderson. The race was christened after millionaire inventor and corporate giant Vincent Bendix, who put up $15,000 in prize money.

Cochran entered the Bendix for several reasons. The race attracted the best pilots and Cochran wanted to be included. She also realized that victory would net a substantial cash award and bring her the fame she desired. Finally, she was drawn to the challenge of spending long hours in the cockpit, facing dangerous conditions and flying at top speeds.

In a speech she delivered in 1939, Cochran explained what a grueling endeavor the Bendix was. "The public thrills during take offs and thrills again while the planes flash across the finish line," she said, "but they never see what goes on behind the scenes."[41] Preparations continued to the very end. "It's a way with racing pilots," Cochran said, "that usually neither they nor their planes are ready until the last minute. All their well laid plans for sleep and other forms of training go astray. But also they are so busy that they don't get jittery at the last."[42] As Cochran explained, the racers faced much danger. Before take-off, crew members, or perhaps even the pilot, would fill the airplane with several gallons of gasoline. The fuel had two major effects: (1) It weighs down the airplane and makes take-off difficult; and (2) It creates a fire hazard, making the airplane a potential fireball in a crash. According to Cochran, "The pilot may be wondering whether he will reach the end of the field alive."[43] The race itself was physically exhausting, lasting some eight to ten hours. As Cochran noted, the public "see[s] the pilot again smiling at the end of the race, but they are usually not close enough to see the deep lines of fatigue and worry in his face."[44] The Bendix caused Cochran so much stress that she lost six pounds per race. Other pilots, like Roscoe Turner, had terrifying nightmares about crashes and fires. The night before the 1935 Bendix Race, Turner had lain awake worrying about what might happen if a wing fell off his airplane.[45]

Cochran's first two Bendix Races were unsuccessful. In 1935, she dropped out of the race because of mechanical failure. She entered with the Northrop Gamma she had tried to fly in the MacRobertson, but now rebuilt with an air-cooled Pratt and Whitney engine. Much to her disappointment, the Gamma had a bad vibration that the manufacturers could not fix. Fearing for their reputations, the Northrop and Pratt Whitney people asked Cochran to reconsider flying the airplane. If the Northrop Gamma crashed, both manufacturers would suffer bad publicity.[46] Cochran initially insisted on going ahead and made a shaky, yet successful, take-off. During the race, however, the engine overheated and the airplane vibrated badly. Somewhere around the Grand Canyon, Cochran saw an electrical storm up ahead. Fearing the storm and knowing that for the next several hundred miles there would be no place to land, she decided to dump her fuel and land at the nearest airport, in Kingman, Arizona.[47]

In the 1937 Bendix, Cochran faced 15 opponents; two of whom were flying military-type aircraft.[48] Frank Fuller and his chief test pilot, Frank Sinclair, entered the race with Seversky P-35s. Cochran raced with a 600-horsepower, green and orange Beechcraft. She finished third, her best showing yet, and also collected the $2,500 purse awarded to the first woman to finish. Fuller and his Seversky had won, and Cochran knew that in order to win in 1938, she would need a Seversky of her own. Cochran got her wish after testing one of Seversky's controversial airplanes.

Seversky had a lot riding on the 1938 Bendix. If his airplanes

Jacqueline Cochran is greeted by Vincent Bendix in 1938 after winning the Bendix Race in a Seversky P-35 (photograph courtesy of Smithsonian Air and Space Museum, SI-79-3161).

could perform as well as they were designed, he hoped the Army would reverse its earlier decision. Seversky had handpicked Jacqueline Cochran to show the world what a Seversky pursuit plane could do. If she could out-fly her nine male competitors, she might dispel the doubts of the Army Air Corps officials. She would also bring attention to the skill of women pilots.

In July, Cochran began preparations for the Bendix. She had high hopes of completing both legs of the race. After crossing the finish line, the Bendix racers had the option of remaining at the Cleveland Airport or flying over Bendix, New Jersey, and landing in New York, at a chance for more money. Cochran hoped to capture the women's transcontinental record and the additional $7,000.

The 1938 Bendix attracted ten contenders, nine men and Jacqueline Cochran. As the only female entry, Cochran knew she was in the spotlight — her every move scrutinized by the press. She also realized that her performance in the race would reflect on other female flyers. Frank Fuller, who flew his Seversky to victory in the 1937 race, was the favorite. While the press noted that "Fuller will have to fly his heart out to win," they continued to refer to him as "the man to beat." Other contenders included: Paul Mantz flying a Lockheed Orion; Max Constant, Ross Hadley and Robert Perlich in Beechcrafts; George Armistead in Cochran's rebuilt Gee Bee Q.E.D.; Lee Gelback in a Wendell-Williams; Frank Cordova in a Bellanca 28-92; and John Hichney and Charles LaJotte in a Spartan 7W monoplane. Cochran, in the other Seversky, was expected to take second place.[49]

Cochran was pleased with her take-off and relieved that she had left the airport before the morning fog rolled in. She consulted her four preplanned routes and decided on course number two, which was approximately 100 miles north of the TWA flight course to Kansas City, crossing Santa Fe, New Mexico. "The first hour went fine," she recalled. Then "over Arizona I hit 'the soup.'"[50] A heavy fog impeded her visibility and interfered with her radio reception. From Arizona to Cleveland, she flew the Seversky solely by instruments, which she found "very tiring after the first couple of hours."[51] As her eyes danced from one instrument to another, she attempted to keep her altitude steady and the airplane's wings even. She tried flying above the weather, but at 23,000 feet she was met by ice and freezing temperatures. "I was perishing with cold," she recalled. "My feet felt like chunks of ice."[52] The high altitude blew the top from her thermos of hot coffee, turning it cold. Breathing was another problem. Since oxygen masks were nonexistent, she breathed oxygen from a pipe stem between her teeth that was attached to an oxygen tank. And while contending with the cold and lack of visibility, she noticed that the plane was becoming "wing heavy" or one wing was not feeding out the gas with its counterpart. At "about 23,000 feet with my windshield coated over the ice ... the engine stopped."[53]

To restart the engine, Cochran had to bend over to reach the switch. When she raised up she discovered the airplane had tipped over on one wing, practically on its side. In an effort to regain control, Cochran sent the plane into a dive toward the ground. "I don't worry about being killed," she once commented, "but I don't want to be wrongfully remembered."[54] Cochran lost several thousand feet of altitude before she straightened up and evened out the wings by flying at an angle with the empty wing lower than the full one. Slightly off course, Cochran roared into Cleveland, flying so fast she missed the airport. "It was Corrigan day at the airport," commented Cochran as she explained how she flew in from the wrong side, "so it was the right thing to do anyway."[55] Overflying the airport or not, Cochran had arrived, the third to leave Los Angeles and the first pilot to land in Cleveland.

The racing officials greeted Cochran and accompanied her to the grandstand. For 30 minutes, she anxiously endured their company. As soon as she could escape the ceremonies, Cochran took off for the second leg of the race to New York. When Cochran reached New York, she discovered she had indeed won the Bendix and set a new woman's record. Her time between Burbank, California, and Bendix, New Jersey, was recorded as 10 hours, seven minutes and 10 seconds. Cochran had beaten Laura Ingalls's 1935 record by some three-and-a-half hours.[56]

Jacqueline Cochran had flown the Bendix course in eight hours, ten minutes, and 31 seconds. She had beaten favorite Frank Fuller by 18 minutes. "It was the toughest weather the Bendix Racers [ever] encountered," commented Cochran, "and yet luck played with them."[57] Fuller blamed the weather for his loss. With no radio and zero visibility he said he flew mostly "by the seat of his pants."[58] According to Paul Mantz, "He did not see the ground between Albuquerque and Wichita."[59]

Jacqueline Cochran

For Cochran, the Bendix Races ushered in new opportunities. She gained national recognition and became a respected voice for the cause of aviation. Cochran's 1938 victory, coupled with Earhart's disappearance the year before, helped make her America's number one woman pilot. She used her position to further the cause of aviation as well as women aviators. Cochran also encouraged women who did not choose to fly to become air minded and take advantage of the airlines.[60] Cochran wanted to stimulate the aviation industry. Like most women pilots, she was willing to sacrifice the cause of feminism for the cause of aviation. Cochran believed her participation in the Bendix Races helped women pilots carve out a role in World War II. She willingly took risks in an effort to improve the image of women flyers and make herself a better pilot.

In 1948, Cochran listed winning the 1938 Bendix Race as one of her greatest feats in aviation.[61] A mere six years after earning her pilot's license, Cochran found herself making aviation history. Throughout the 1930s, her many aviation accomplishments helped blaze a trail for others to follow.

* * *

Jacqueline Cochran was the best woman pilot of her day — even better than Amelia Earhart. She could barnstorm, fly the Gee Bee Racer (the world's most dangerous airplane), win air races, set speed records, and conduct high-altitude experiments. In recognition of her skills, Cochran was presented with several aviation awards, including the General William E. Mitchell Memorial Plaque in 1938, and the Harmon Trophy in both 1938 and 1939.[62] In her lifetime, Cochran won 15 Harmons, international trophies which recognized her as the world's outstanding woman flyer.

As a successful speed and air racer, Cochran gained great notoriety. She was recognized worldwide, her name and photograph appearing in newspapers such as the *New York Times* and in magazines. A Jacqueline Cochran-inspired cartoon even appeared sometime after her Bendix victory. In 1939, she was named the first woman to serve on the Collier Trophy Selection Committee, which each year honored advances in aviation.

Cochran was one of a handful of talented women flyers in the 1930s.

3. America's Number One Woman Flier

Amelia Earhart was the best known. Other women pilots included speed racer Louise Thaden, socialites Anne Morrow Lindbergh (wife of the famous Charles Lindbergh) and Nancy Harkness Love, and Helen Richey, the first woman airline co-pilot. In all, about 500 women were pilots in the 1930s, less than one-thirtieth of all aviators.[63]

In 1929, when women began participating in air races, they faced many obstacles. Air racing was a dangerous endeavor and women were often barred from participation. If allowed to race, they were expected to race against each other. The Bendix was a good example of the gender-biased nature of air races. Prior to 1935, women pilots could not compete. In 1933, Cliff Henderson, the organizer of the Bendix, bluntly stated: "Women have no more place in the National Air Races than on the automobile racetrack at Indianapolis."[64] Apparently, Henderson's decision was influenced by the tragic death of Florence Klingensmith, a 23-year-old pilot who had lost her life in a Gee Bee in the Frank Phillips Trophy Race that same year.[65] Henderson feared a similar incident could taint the Bendix. To Henderson, a dead pilot was a tragedy; a dead woman was a liability.

Women pilots soon discovered advantages and disadvantages in competing with men. At national air races, which offered big money and challenging competition, reporters often spotlighted women flyers over their male counterparts. Even Roscoe Turner, the popular pilot who flew with a lion cub, had a tough time competing with Amelia Earhart. Women, however, were often at a technological disadvantage with slower aircraft than those flown by men. Male pilots often entered races with specially built racing planes, while women entered with stock model planes.[66] A frustrated and angry Amelia Earhart complained to Louise Thaden: "Manufacturers refuse us planes, [and] the public has no confidence in our ability. If we had access to the equipment and training men have, we could certainly do as well."[67]

Henderson's decision to forbid women pilots from competing marked the second time in 1934 that women had been snubbed. In February, male pilots had formed the Professional Racing Pilots' Chapter of the National Aeronautic Association but refused to allow women to join. Such actions made women pilots more determined to be recognized. Unable to participate in the 1934 Bendix, women pilots banded together

47

and held their own race, the Women's National Air Meet in Dayton, Ohio. Dayton businessmen donated $3,000 in prize money. "A dab of money compared to the purse at the National Air Meet," wrote Charles Planck.[68]

Some women pilots, however, favored the all-female races. As Susan Ware noted, "Similar to the rationale of single-sex education, [women pilots] believed that the absence of men allowed women to excel and build the qualities necessary to compete in the wider aviation world."[69] Others, most notably Amelia Earhart, objected to how women were treated.

Throughout her life Earhart tried to fight discrimination and break down gender barriers. Her outspoken character was shaped by her commitment to liberal feminism, which she expressed in her ambitions for women, her carefree personal appearance, and her modern marriage, wherein she was considered a partner rather than a submissive wife. As an aviator, Earhart fought for equal treatment for women flyers, especially in air racing. At the first Women's Air Derby, Earhart criticized the all-male organizing committee for requiring contestants to have a current license and 100 hours of solo flying time. Earhart estimated that only 30 women could meet those qualifications, and only 20 women would enter the race.[70] The committee also wanted the women to begin the race in Omaha rather than California in order to spare the women "the danger of crossing the Rocky Mountains."[71] When all 20 contestants refused to fly unless the course remained unchanged, the committee reconsidered.[72] Earhart also openly protested the first Bendix Air Race. When Cliff Henderson forbade women flyers, she refused to fly movie star Mary Pickford to the race after Pickford had agreed to open the ceremonies for the 1934 Race.[73]

Air races often maintained perceptions of female inferiority. Much to Earhart's displeasure, the Women's Air Derby was termed the "powder puff derby." Furthermore, the contestants were referred to as "Ladybirds," "Angels" or "Sweethearts of the Air." "We are still trying to get ourselves called just 'pilots,'" commented Earhart in 1932.[74] One article described the race as a mixture of "death, romance, fever, sunburn, lipsticks and powder puffs."[75] The women were also forced to endure being wined and dined at every stop. According to one participant, "It [the race] wouldn't be so bad, if it were not for the checking and the chicken."[76] The "checking" was a reference to the work pilots put in before flying, including laying a course and consulting, sometimes even redrawing, their maps. While the women tired of dining on chicken, they patiently endured every banquet. According to Joseph Corn, "By acquiescing in the social whirl, however, no matter how well intentioned, the Derby pilots inevitably gave support to those who would claim women flyers were frivolous and unprofessional."[77]

One deadly mishap of the race reinforced such views. Tragically, Marvel Crosson died when she became ill and was thrown from her airplane. Her death raised questions about women in air races. A writer for the *New York American* noted, "Air racing for women should be discouraged as a far too hazardous adventure."[78] Nevertheless, 16 women finished the race, which prompted Amelia Earhart to note that the Women's Air Derby had "the highest per cent of 'finishers' in any cross country derby, up to that time, for men or women."[79]

Jacqueline Cochran was not a liberal feminist like her friend Amelia Earhart. In fact, she disdained feminism, believing such ideas conflicted with true womanhood. Cochran was very much a rural Southerner with traditional beliefs about marriage, family, and women's place in society. While Cochran strongly believed in women's abilities, she felt their first duty should be to the home, a belief that conflicted with her own life. "I don't know why people worry about woman's place," she later explained in an interview. "Women have the same intelligence as men, they have the same ability. But they should not forget that they're women. They should populate the world; they should have a woman's place in the world."[80]

Cochran's opinions were molded by her past. Even after she obtained wealth and celebrity status, a part of her remained the impoverished orphan Bessie Pittman. "Whatever I am is elemental," Cochran commented in 1954, "and the beginnings of it all have their roots in Sawdust Road."[81] As a child, Cochran desired a stable, loving family. After she became an adult, Cochran placed great importance on home and family, goals that feminism seemed to betray. In 1948, after many awards, honors, and aviation firsts, Cochran said her greatest achievement was "having a successful marriage."[82]

Cochran also rejected feminism because, in her view, it conflicted with her two great passions, aviation and the beauty business. Women pilots, Cochran felt, would be more acceptable if they conformed to gender stereotypes rather than espousing feminism. She also used her gender to protect the cause of aviation. At the same time Cochran entered aviation, she was establishing Jacqueline Cochran Cosmetics. From a business point of view, a glamorous woman with a flair for make-up and fashion would be more appealing than a feminist eschewing stylish clothing and "feminine look."

Despite her beliefs, Cochran did open some doors for women in aviation. At times, Cochran's motives were purely selfish. She chose her causes carefully and often fought to benefit herself more than others. Still, Cochran wanted women to have a place in aviation.

Cochran challenged gender stereotypes just by earning a pilot's license, an accomplishment achieved by few women. She soon proved that a woman could race, participate in aviation test research and fly as well as, if not better than, a man. While Cochran disliked feminism, she was an outspoken woman with a relentless desire to do things her own way. At times, Cochran would argue for fair treatment for women pilots, and when she did, she was often successful.

In 1935, the Bendix was opened to women, partly because of Jacqueline Cochran. Although no activist, Cochran wanted to race, and petitioning the Bendix committee was her only recourse. "I got hold of Mr. Bendix," recalled Cochran years later, "and I said, I can fly as well as any man you are flying in these races."[83] In 1935, the Bendix Racing Committee reversed its decision and allowed women to enter the race after Cochran had several heated discussions with racing authorities. She hammered out an agreement in which women would be allowed to enter the race, provided Cochran could obtain waivers from the male pilots.[84] Essentially, the officials required female pilots to seek permission to race from their male counterparts. Whatever the women pilots may have thought about the humiliating agreement in private, they accepted the deal. As Susan Ware had noted, they appeared to handle this indignity as they had always done, by "ignoring it and just going about their business."[85]

Like Earhart, Cochran was often annoyed by the patronizing attitudes toward women pilots. For example, after Cochran won the 1938 Bendix, beating nine men, her victory was met with skepticism and surprise that would never have been accorded a male pilot. When she landed in Cleveland, a newscaster, upon learning that Cochran had reached a maximum speed of 340 miles an hour, said: "Say, that's pretty fast for a woman!"[86] Cochran shot back: "For a woman! Mr. Uttal, do you think speed flying is for men only?"[87] Uttal replied, "But you must have had ideal flying conditions to have made such fast time."[88] Cochran explained that "conditions were really terrible" and she had flown instruments at least two-thirds of the time.[89] The press also commented on Cochran's hair and figure and pointed out how she reapplied her lipstick and asked for her husband after landing in Cleveland.[90] The *Dallas Times-Herald* reported that Cochran "prettied herself in her ship while awaiting completion of the refueling process."[91] Another article, subtitled, "Successor to Earhart is personable Redhead; loves Husband," included a description of Cochran's "powder puff" living room and assured the public that "she's feminine."[92]

As an added insult, some skeptics questioned Cochran's victory. Some time after Cochran's appearance at the Thompson Company Aviation Ball, a rumor spread that a male pilot had actually flown in Cochran's place. After all, no "mere woman could have done what she did," and no woman could look so fresh after an eight-hour flight.[93] The rumor was linked to the Army Air Corps brass who may have resented the boost Cochran gave to Seversky and his plane.[94]

In her own way, Cochran fought against stereotypes and at the same time, perpetuated them. In the air races, as in her later career, Cochran presented a dual image. She was a very feminine, glamorous woman with a flair for challenges and a willingness to take risks. A tall, attractive blonde, Cochran emphasized her appearance. When flying, she favored tailored suits and long flowing scarves, and she carried make-up. She stood in direct contrast to Amelia Earhart. During the 1934 MacRobertson race when Cochran and Smith landed in Romania, her biggest concern was her appearance. When Cochran stepped out to greet the Romanian Air Minister Radu Irimescu, anxious reporters tried to photograph her in her coverall flying suit. Cochran protested and convinced the newsmen to give her time to comb her hair and apply rouge and lipstick.⁹⁵ While her habit of applying make-up annoyed some of her female counterparts, Cochran felt no remorse. Since she was starting a cosmetics company, she needed to maintain a glamorous image. What better way to advertise her products than to present herself as glamorous? Cochran may also have used glamour to further the cause of aviation. "If [Cochran] looked fresh and glamorous after a flight, it would show flying as being smooth and restful."⁹⁶ At the same time, she may have hurt the cause of women aviators. Instead of proving that women pilots were as competent as men, Cochran reinforced the idea that women were frivolous.

Cochran sometimes resorted to her femininity to advance the cause of aviation. On several occasions she took the blame for failures, portraying herself as a weak woman, rather than bring negative publicity to an airplane or its manufacturer. Cochran put aviation first and the advancement of women aviators second. She certainly did this in 1937 when she allowed Alexander de Seversky to use her gender to promote the P-35.

Cochran also used her gender to protect manufacturers. When reporters interviewed Cochran about why she quit the 1935 Bendix, she failed to mention the Northrop Gamma's faulty engine. Instead, she simply told them that she quit because she was too tired to continue. "I saw no reason," she said, "to give an alibi which would put the blame on the plane or engine."⁹⁷ Cochran lied about the Bendix to protect the Prat and Whitney and Northrop Companies, as well as the cause of aviation. Unfortunately, she hurt the reputation of women pilots. By feigning exhaustion, Cochran seemed to confirm that women were too frail and delicate to participate in grueling national air races. She believed if she conformed to female stereotypes, she would help the cause of aviation.⁹⁸ Cochran was also quite willing to use her gender to promote the safety of flying. While she declined to crusade for woman, Cochran was very concerned about the future of aviation. "In my own cause," she said, "the risks I would consider would not be death as much as burns and painful disfiguring injuries or damages to a valuable plane or harm to the cause of aviation."⁹⁹

* * *

By the end of the decade, Cochran emerged as an internationally known aviator. She entered several air races and became one of the few

3. America's Number One Woman Flier

women to win the prestigious Bendix. Cochran also contributed to advancements in aviation by testing airplanes for speed, and pilots for endurance. At the same time, she was winning races and breaking speed records Cochran was establishing a cosmetics company, adjusting to married life and dealing with the disappearance of her close friend Amelia Earhart.

In the 1930s, Cochran was still new to aviation, having earned a pilot's license in 1932. Despite this fact, she was at her finest in 1938. At no other time did she receive more praise for her skills as an aviator. Cochran was a "golden girl" in the golden age of aviation. During World War II, she gained a reputation for bossiness and egocentrism, which seemed to overshadow her flying skills. After the war, she found herself living in an era where aviation accomplishments were no longer thrilling. In many ways, the 1930s were the best years of her life. Cochran found where she belonged — in the air.

4

An Unusual Life

Between 1932 and 1939, Jacqueline Cochran had a dramatic rise to fame. She quickly became a skilled flyer and made a name for herself breaking records, speed racing, and testing airplanes. After winning the 1938 Bendix Air Race, she was recognized as the world's best woman flyer.

Still, Cochran presented a paradoxical image. She out-flew both men and women pilots and maintained a glamorous, almost movie star-like image. Meanwhile, as she opened doors for women flyers, she willingly perpetuated unfavorable perceptions and stereotypes of women to promote aviation. Cochran's ambivalent and contradictory view of women can be seen in her marriage, business career and relationship with Amelia Earhart.

* * *

Jacqueline Cochran's quest for independence and freedom was realized in the mid–1930s. Despite her lack of education and poverty ridden childhood, Cochran became a successful businesswoman, as well as an internationally known aviator. Her willingness to work hard, combined with her keen mind, made up for her lack of formal education. These factors, however, do not adequately explain her success. She started Jacqueline Cochran Cosmetics in 1935, in the middle of the Great Depression, when businesses were failing and many Americans suffered financially. Her success depended heavily on her association with Floyd Odlum. By marrying a tycoon, she had access to business advice and capital as well as financial support for her aviation ventures.

Cochran owed much to Floyd Odlum. Ever since the two had begun dating in 1932, it is likely that Odlum provided her with both funds and strategy. It is unlikely that she could have built a successful business without his help. In 1936, Floyd Odlum and Jacqueline Cochran were married, after dating secretly for four years. Cochran explained that the relationship "had begun quietly, privately, because he didn't want to hurt his sons, Stanley or Bruce."[1] At the time, Odlum was involved in a failed marriage. Cochran never explained why it took Odlum three years to finalize a divorce. When Odlum and Hortense McQuarrie did part, it appeared to be an uncontested divorce.[2]

At first glance, the quiet, reserved and introspective Odlum seemed an odd match for the bold, outspoken Cochran. While she pursued flying, he was content to stay at home and mold clay figures. Certainly, the two came from different worlds. Odlum was born into a lower working-class Michigan family and worked his way through college. Meanwhile, Cochran grew up in poverty with no chance for an education. "Temperamentally we may have been miles apart," commented Cochran on her relationship with Odlum, "but when it came to knowing what we wanted out of life — security, power, and a certain kind of fame — we were very much alike."[3]

This quest for security had always been an important goal in Cochran's life. Born into poverty, she drove herself to establish a better life. The memories of her harsh childhood were never far from her mind. Marrying Odlum in 1936 helped her to achieve that life. "I think what she got when she married Floyd," commented test pilot Chuck Yeager, "was an opening door; because what she saw was the opportunity not only to have drive and ambition but to use power."[4] Both Cochran and Odlum were self-made people. They left their small towns and reached for great heights. They both possessed keen business minds and enjoyed running companies. Cochran and Odlum were perfect companions. They built a 40-year marriage based on love and mutual admiration. To Cochran, Odlum was "the gentlest, kindest, and most generous human being" she had ever known.[5] Meanwhile, Odlum referred to Cochran as "the most interesting person I ever met."[6] He admired her loyalty and the fact that she was her own person. According to Oldum, Cochran's unique personality was "backed by imagination, mental alertness, energy and power of expression which act as a spark plug for interesting living."[7]

After their wedding, the two moved to California and built the Cochran-Odlum Ranch. There they entertained distinguished visitors such as Howard Hughes, Amelia Earhart, and Dwight D. Eisenhower. Although she was proud of her husband, Cochran did not like being called Mrs. Floyd Odlum, and continued to use her chosen name. She would not sacrifice her hard-earned identity even in marriage. When she chose the name Jacqueline Cochran, she had felt a sense of independence, a break from her past. To surrender the name meant giving up a part of herself.

Floyd Odlum seemed to admire Cochran for her independence and ambition. Despite his wealth and power, in many ways he lived through her. Beset with arthritis and perhaps his own introverted personality, Odlum was content to be the man behind the woman. "Floyd always encouraged Jackie," commented Helen Lemay (wife of Air Force General Curtis Lemay), "and was so proud of her records, her accomplishments, where she had come from, and all that."[8] Cochran had fond memories of Odlum lying on their couch planning her racing strategy, mapping "routes and measuring engine possibilities against fuel-consumption tests."[9] Both before and after their marriage, Odlum encouraged Cochran's flying. Before hard races, he would present her with a yellow rosebud.[10] Odlum often worked backstage, helping Cochran comply with entry rules and in locating the right engine or equipment. He was ready to respond when Cochran needed his inspiration prior to taking off in the 1935 Bendix. "That night I needed to hear his calm voice of reason more than anything," recalled Cochran. "And he was there for me."[11] Odlum was the first to greet her after her 1938 Bendix victory.

Floyd Odlum also took Cochran's idiosyncrasies in stride. Cochran admitted to being very superstitious, particularly about the number 13. Strangely, she did not seek to avoid the number, but considered it lucky. Cochran used "13" as her official racing number and proudly painted it across several of her airplanes. The original cover of her 1954 autobiography showed 13 pictures of Cochran. (Ironically, 13 mourners attended her 1980 burial.) Cochran also believed she possessed psychic powers, something she termed as a "real sixth sense or intuition ... [that] apparently burst into full bloom as extrasensory ability."[12] At one point, the Odlums invited a friend who claimed to be a medium to their home, who supposedly levitated furniture and performed psychic experiments.[13]

4. An Unusual Life

Together, Odlum and Cochran reveled in their business successes and her aviation triumphs. They also dealt with the tragedies of Cochran's two miscarriages and Odlum's worsening arthritis. The Odlums appeared to have a strong marriage, built on trust and partnership. She had grand ideas, and Floyd Odlum had the money to finance them. When Cochran wanted a cosmetics company, Odlum helped her start a business.

Few records exist concerning Jacqueline Cochran Cosmetics. Before shipping her papers to the Dwight D. Eisenhower Presidential Library, Cochran removed most of the files concerning her personal finances and business activities. Archivists at the Eisenhower Library assume the records were destroyed.[14] Apparently Cochran wanted to conceal the extent of Odlum's assistance.

* * *

In the 1930s, as Jacqueline Cochran was building a career in the male-dominated field of aviation, she was also pursuing the beauty profession. Defying the era's standards, she built a profitable and successful cosmetics business, becoming one of the era's few businesswomen. Cochran used her glamorous appearance and take-charge personality to her advantage. She was a pioneer in the field of beauty, as well as aviation.

Despite the Great Depression, American women continued to work in the 1930s, but most worked in domestic or clerical positions that men would not accept.[15] Fewer women owned their own businesses, and still fewer were successful. Those who did, however, tended to be in "women's" lines, such as cosmetics.[16] The beauty industry had emerged in the 1890s and began to flourish in the more affluent 1920s, when women began perming their hair and wearing make-up.[17] The beauty profession soon became an acceptable and popular career for women, even in the economically troubled 1930s. Despite the depression, new companies like Almay, Clairol, and Maybelline were established and new beauty products appeared on the market.[18] In 1935, *Fortune* named Elizabeth Arden and Helen Rubinstein, businesswomen who sold cosmetics, two of the most successful women in America.[19] Cochran, a natural beauty, seemed well suited for the cosmetics profession. Soon she too became an accomplished businesswoman, despite her lack of education and money. Unfortunately, Cochran left few records concerning her business, including the extent of Floyd Odlum's financial backing.

Cochran, of course, had worked in the beauty business for many years. In fact, she had been involved in the business since she was 11 years old.[20] Perhaps Cochran was attracted to beauty to feel better about herself. After all, she grew up as a scraggly little girl wearing feed sacks and rarely bathing. She also seemed to have enjoyed her ability as a stylist to make other people happy.

After quitting Antoine's in 1932, Cochran decided to re-enter the beauty profession in 1934, and she established Jacqueline Cochran Cosmetics on Fifth Avenue in 1935. The new firm fed her desire for independence, and allowed her to create her own products. "I wanted no part of other people's products ... because I was crazy enough to think I could do better."[21] With a laboratory in Roselle, New Jersey, the company began creating original creams and fragrances and revolutionizing the world of beauty.[22] Cochran was responsible for many "firsts" in the cosmetics industry, especially regarding hair dyes. Working from "scratch," she mixed peroxide and aniline dyes to create new hair colors, including topaz, chestnut, and blonde.[23] Some of her more popular products included the Flowing Velvet facial cream and a stackable make-up kit called the "Perk-Up Stick." This was a purse-sized black tube that included a cleansing cream, foundation, night cream, cream rouge and face powder.

Apparently Floyd Odlum's 1932 suggestion that Cochran needed wings to build a successful business during a depression was sound advice. Despite a shaky economic climate, her business became quite profitable. At Jacqueline Cosmetics, Cochran served as owner, inventor, and saleswoman, combining flying with peddling her products. As she recalled, "I'd go on the road and in the air, to sell. The combination of my flying career and cosmetics worked. In fact, my first two department store accounts — Pogue's in Cincinnati and Halle Brothers in Cleveland — ordered precisely because of the flying angle."[24] By 1938, Cochran owned prestigious shops in Chicago, Lake Forrest, Illinois, and Los Angeles.

Some of the problems Cochran encountered as a pilot led to innovations in beauty. Disturbed by the condition of her skin after high altitude and long-distance flights, Cochran set out to create the perfect moisturizer. Her days were very mixed, she explained. "I'd put in my time up front on Fifth Avenue, on the road selling, and I would putter in the lab trying to refine a product."[25] The result of Cochran's hard work was a product called Flowing Velvet, a lemon-colored moisturizer famous for its soft, dewy feel. She also helped revolutionize lipstick, and some of her products made it to Hollywood. For example, Cochran created Marilyn Monroe's lipstick and gloss for the 1953 movie *Gentlemen Prefer Blondes*. "There was a lot of formula breakthroughs for which Jacqueline Cochran was responsible," commented Scott Vale, Vice-President of Training at Jacqueline Cochran Cosmetics in 1985.[26]

Combining flying and beauty allowed Cochran to pursue her two great loves. Her company's slogan "Wings to Beauty" reflected her dual pursuits. With the slogan painted across the side of her airplane, Cochran flew from store to store, selling and publicizing her products. In 1938, she flew some 3,000 miles in one week of business.[27]

Cochran was proud of her company and rarely missed a chance to publicize it during interviews. She played up her femininity and seemed to enjoy being referred to as a "glamour girl." Cochran realized that her livelihood depended on cultivating a certain type of feminine look, a look quite different from that of Amelia Earhart. Fashionable and stylish, Cochran used herself as an advertisement for her products. She wore the beauty profession as a badge of honor, proving herself a creative innovator and savvy businesswoman.

After World War II, Cochran resumed her cosmetics career and expanded her company. In 1945, she bought a six-story building to accommodate her growing business.[28] Three years later, Cochran formed a new corporation called Jacqueline Cochran, Inc., which replaced Jacqueline Cochran Cosmetics.[29] In both 1953 and 1954 Cochran was voted "Woman of the Year in Business," in Associated Press polls. She sold her cosmetics company in 1963.[30]

* * *

Despite their different attitudes and beliefs, however, Cochran developed a strong friendship with her contemporary Amelia Earhart. Cochran referred to her as a "great flyer" and "my inspiration."[31] She had great respect for Earhart's originality, courage and dedication. While she acknowledged their differences, Cochran praised Earhart for her beliefs: "She was an ardent feminist [who] did much to interest women in flying and to advance their cause."[32]

Cochran and Earhart were two very different pilots, pursuing different goals for advancement in aviation. While Cochran made a name for herself in the area of speed racing, Earhart pursued distance flying. She began her aviation career by flying across the Atlantic, as a passenger. Later, she made more noteworthy accomplishments, including soloing the Atlantic while piloting the plane, and soloing between Hawaii and California.

Despite her accomplishments, Earhart was never an "ace flyer" like Cochran. According to Earhart biographer Susan Ware, "While Amelia Earhart may have been the most well-known woman flier at the time, she was not yet the most experienced or competent."[33] Earhart took many years to learn her flying skills, while Cochran was a natural flyer, soaring after only three weeks of lessons. Still, Cochran was never resented by Earhart.

Cochran first met Amelia Earhart in 1935. The two were dinner guests of Paul Hammond, a mutual friend. According to Cochran, within five minutes of meeting Amelia she knew they would become fast friends.[34] Soon, they planned a flight from New York to California in the same Lockheed Electra airplane in which Earhart would disappear. When bad weather prolonged the trip to an unexpected six days, they spent much of their time discussing extrasensory perception. Later, they practiced finding Amelia, believing they had discovered a "safety net" in case she should ever disappear.[35]

During the last two years of Earhart's life, she and Cochran enjoyed a close friendship. Earhart was a frequent guest at the Cochran-Odlum Ranch, where she found a respite from her fame. A 1930s photograph captures the bond between the two. The pictures show Cochran and Earhart sitting on the diving board at the ranch's swimming pool. Cochran, in her slacks, matching jacket and flowing scarf, sits with her feet dangling above the water, holding a cigarette as she gazes into the pool. Earhart, who is dressed in casual pants, a plaid button-up shirt and leather jacket, is curled up on the end of the board, smiling slightly as she gazes to her left. Neither woman is looking at the camera. Side by side, they seem like old friends, simply enjoying each other's company. According to Cochran, "During the last year of her life, I was closer to Amelia than anyone else, even her husband, George Putnam."[36] According to Earhart biographer Susan Ware, Cochran "developed a deep personal rapport with Earhart."[37]

While both Cochran and Earhart were independent, ambitious women carving out fame in a man's world, they were more different than alike. Earhart grew up in a small Kansas town, raised by a close-knit family. The Earhart family "lived in reasonable comfort," Cochran in poverty.[38] Certainly, each had her own personal style. Earhart was a dedicated feminist, although she preferred the term "modern thinking."[39] According to one biography, "Her thinking, her entire philosophy, was based on her certain knowledge that she was the equal of anyone."[40] Earhart supported the women's suffrage movement, spoke out for birth control and advocated drafting women for war. Ignoring the dress code of her era, Earhart disdained ruffles and make-up and chose to wear her hair short and curly. "She was as simple as she was famous," commented Cochran of Earhart. "She preferred a pair of slacks to silks."[41] Meanwhile, Cochran played up her femininity with carefully arranged golden curls and personally designed lipstick. She also advocated limited roles for women in society. Cochran believed that mothers should not work outside the home and was especially against women in combat: "When women start [combat] flying, then something really ghastly will have happened to the world."[42]

One similarity, beyond love of flying, was the women's married lives. Both married successful, supportive men who encouraged and promoted their wives' careers. Earhart's husband, George Putnam, was largely responsible for his wife's fame. He widely publicized her ventures and helped make her a better pilot. Floyd Odlum was a less intrusive figure in Cochran's career, but he often wrote letters on her behalf and helped her prepare for races and flights. Odlum also provided her with financial support. Yet, even in marriage, Cochran and Earhart displayed marked differences. Cochran dearly loved and depended on Floyd Odlum. Earhart, according to fellow aviator Fay Gillis Wells, described her union with Putnam as "a marriage of convenience."[43] On the eve of her wedding, in fact, Earhart had written a letter to Putnam assuring him that she would not hold him "to any medieval code of faithfulness" nor should he expect her to abide by this code.[44]

Both Putnam and Odlum offered their respective wives the kind of marriage they needed. They understood that flying was much more than a career to Cochran and Earhart; it was a vital part of their lives. Flying made them who they were. Both Putnam and Odlum took pleasure in their wives' successes and, because of their own fortunes, they could offer their wives unlimited financial support. Neither Odlum nor Putnam was looking for a housewife nor (at least in Putnam's case) a mother for his children. They expected nothing more from their wives than to be themselves. Both Cochran and Earhart were grateful for their husbands' attitudes. Their sense of freedom and independence gave their restless spirits room to soar.

One of Cochran's fondest memories of Amelia Earhart was her unexpected visit following the failure of her first around-the-world flight. Cochran referred to the night as "one of the nicest and most touching evenings we ever spent together."[45] Whether it was psychic foreboding or simple concern for her friend, Cochran harbored bad feelings about another flight.[46]

A few weeks later Earhart returned to the ranch, clutching the plans for her next flight. Cochran studied the plans with her as she worked out the final details. "But I was still uneasy about it all," recalled Cochran.[47] "My sixth sense caused me to have vague apprehensions."[48] Out of concern for her friend, Cochran packed Earhart a survival kit that included fishhooks, lines, an all-purpose knife and the same brightly colored kite Cochran had taken on her London-to-Melbourne race. According to Cochran, she personally obtained a three-week supply of canned powdered bananas and had them placed on Amelia's plane.[49] Cochran also provided Earhart with financial support. Earhart's last flight was partially financed by Odlum and Cochran and Vincent Bendix. The money, however, was irrelevant to Cochran. She worried about her friend.[50]

On July 2, 1937, Amelia Earhart and navigator Fred Noonan took off from Lae for Howland Island. Later that day, Cochran received the word that Amelia was missing. Unable to locate the island, Earhart and Noonan disappeared somewhere over the Pacific Ocean. Desperate for leads, a shaken George Putnam asked Cochran for help. He had always been skeptical, even mocking, of Cochran's psychic abilities, but now he grasped at any chance of finding Amelia.

According to Cochran, she did experience a vision concerning Amelia's whereabouts. Right after Earhart's disappearance, Cochran scribbled down some key information, which Putnam supposedly pocketed and never revealed. Cochran claimed to "see" the Lockheed Electra floating just northwest of Howland Island. Amelia was unhurt, but Noonan was unconscious after having injured his head on the bulkhead during the crash. Cochran also identified an American ship named the *Itasca* and a Japanese fishing boat called the *Maru* or perhaps the *Mari*. The *Itasca* had been in contact with Earhart shortly before her airplane disappeared. Cochran claimed she had never heard of the ship until it appeared in her vision.[51]

For the next three days, Cochran continued to "see" Earhart. Directed by Cochran's visions, Putnam cabled Secretary of Commerce Daniel Roper and requested that the search for Amelia be concentrated around the Ellice, Gilbert and Marshall islands, Ocean Island and areas northeast of these islands.[52]

On the third day, according to Cochran, she called again to tell Putnam that it was too late to rescue Amelia: "She was gone."[53] The search continued for several months, and even though Putnam accepted Amelia's death by October, periodic searches continued for years thereafter.[54]

The fate of Amelia Earhart may never be known, just as no one knows the validity of Cochran's visions. Although some theories claim Earhart lived out her days on a remote southwestern island called Nikumaroro, her disappearance has a simpler explanation. Most likely, Earhart's airplane ran out of fuel and plunged into the Pacific, too deeply hidden to be found. Following the tragic episode, Cochran "consciously and purposefully submerged [her] sixth sensitivities."[55] Guilt ridden and distraught at not finding her friend, Cochran abandoned her psychic experiments. "They proved no use to me or Amelia on July 2, 1937," she recalled. "I never toyed with extrasensory perception again."[56]

Did Jacqueline Cochran experience visions about Earhart's whereabouts? Perhaps, but Cochran herself raised doubts about what she "saw": "I did [locate Amelia], or at least I THINK I did, but can never prove it one way or the other, and besides it was all to no purpose."[57] Another problem with Cochran's claim is her contention that she had never heard of the ship *Itasca*. If Cochran, as she claimed, had been with Amelia when she had finalized her plans, Cochran should have known that the *Itasca* was scheduled to be in the vicinity to help guide the flyers to Howland. Furthermore, this fact was well publicized prior to Earhart's flight.[58]

With Amelia Earhart presumed dead, Jacqueline Cochran became America's most celebrated woman aviator. While both were famous pilots, Cochran accomplished more for aviation than Earhart. Surviving her friend by some 40 years, Cochran certainly had more opportunities for achievement, but has never received as much attention. Earhart's disappearance and George Putnam's publicity machine carved out her place in history. While contemporaries recognized Earhart for her feats in aviation, she was never as popular in life as she was in death.

In the areas of record making and racing, Cochran would have achieved just as much if Earhart had lived. Earhart had no interest in establishing speed records and no need to compete with Cochran. Although Cochran was the better pilot, she often put Amelia Earhart's name ahead of her own. In 1938, Cochran corrected interviewer Frankie Basch who suggested that she (Cochran) had "probably done more to further the feminine angle in aviation than any other woman."[59] "No..." answered Cochran, "I think that Amelia Earhart has done more for women in aviation than any other."[60]

* * *

By the end of the 1930s, as war broke out in Europe, Cochran found herself in an unique position, a celebrity in the revived world of aviation. Like many Americans she turned her attention to the German war machine with its menacing aircraft and wondered what she could do for the allies. "I fretted as to how I could get into it but for some time could do nothing better than knit afghans for the Royal Air Force."[61] Looking for new challenges, Cochran decided to organize women flyers to aid the British.

5

Women in War

By 1930, Jacqueline Cochran was recognized not just as a talented pilot, but an aviation celebrity. Her new status did not diminish her desire for racing or new challenges. Cochran had once resolved to quit racing if she should win the Bendix, but now she looked forward to more races. "Adventures were always just around the corner," Cochran commented of her ambitious spirit, "and I could turn that corner mighty fast."[1] In 1939, the crumbling of world peace propelled Cochran toward her greatest adventure — organizing women pilots to aid the Second World War.

Unlike in World War I, aviation played a critical role in World War II. Strategic bombing and tactical air war created an unprecedented demand for pilots to fly both combat and noncombat missions. When the United States entered the war, its Army Air Corps would need thousands of pilots. Young men by the hundreds learned to fly, but there was always a need for more pilots. Women fliers were one means of meeting this need, but never before had women been allowed to fly for the Air Corps. When the talented stunt pilot Ruth Law volunteered for service in 1917, she was asked to sell war bonds. At the start of World War II, Jacqueline Cochran was America's top woman pilot, and she was not about to spend the next few years selling bonds. From her work as a test pilot and her great knowledge of aviation, Cochran realized how important military and non-military aviation would be to the war. Years before Pearl Harbor, she perceived the need for pilots and how women might help meet that need. Cochran was determined that women should have a role in wartime aviation, and that she should be at the center of American efforts to train and organize women pilots.

As soon as war broke out in Europe, she began campaigning to organize a unit of women flyers for noncombat missions. As long as the United States remained "neutral," most American leaders gave her little serious attention. In Great Britain, however, she found greater interest, and in 1941 she agreed to head a group of American women who would ferry planes for the British Air Transport Auxiliary Service, a branch of the Royal Air Force. These were the first American women to fly military aircraft in time of war. Cochran led the way by flying a bomber across the Atlantic. And, as always, she broke barriers for women while embracing stereotypes of sexual inequality and feminine glamour.

* * *

When Germany invaded Poland on September 1, 1939, it shattered not only Europe's fragile peace but also traditional gender roles. Across Europe, women in the coming months would begin doing "men's work," and in Russia, Germany, and England this included military flying. Russia trained women for combat missions; England and Germany for "service" flying, such as ferrying planes from factories to bases.

British women wishing to aid the Royal Air Force could join a variety of organizations. The Women's Auxiliary Air Force had been formed before Germany attacked Poland.[2] Anticipating war and the need for additional personnel, the British government issued a Royal Warrant on June 28, 1939, creating the organization. Prior to the creation of the WAFFS, women had joined the Women's Voluntary Services or served in the Women's Auxiliary Territorial Service. Created in 1938, the ATS women worked as motor mechanics, draughtswomen, map readers and anti-aircraft gunners. Members of the ATS shot down enemy airplanes and occasionally received medals. In 1941, the ATS were given full military status in the British Army.[3] The WAFFS had gained experience by working with the Royal Air Force and the ATS. Under the command of the RAF, the WAAFs' principal responsibility was to release male pilots for combat. The WAFFs served as electricians, flight mechanics, meteorologists and map tracers. Women with pilot's licenses joined the Air Transport Auxiliary, a branch of the Royal Air Force, formed in September 1939. Here, British women were integrated with male noncombatants and "performed exactly the same duties as the men, had equal pay and rights."[4] These ATA girls ferried airplanes inside the British Isles.[5]

5. Women in War

Jacqueline Cochran took an immediate interest in the war and wanted to be part of it. She was not an "interventionist," and prior to 1939 had paid little attention to world affairs. Her interest, instead, derived from her desire to be at the center of things and her passionate interest in all aspects of aviation. From the very start of the war, she viewed America's participation as inevitable and believed air power would be the key to victory. When the British Air Transport Auxiliary began flying missions in 1939, it attracted Cochran's attention. She followed its progress and soon formulated ideas for an American version, one in which American women would play a role.

Just weeks after Germany's invasion of Poland, Cochran presented her ideas in a letter to no less than Eleanor Roosevelt. She wrote to the First Lady, detailing the value of women pilots. "National defense and preparedness," Cochran explained, "require the organization of all our resources to prevent waste of time in emergency. In the field of aviation the real 'bottleneck' ... is likely to be trained pilots."[6] Women pilots, she suggested, could be used to fly ambulance, courier, commercial or transport airplanes. "This requires organization and not at the time of emergency but in advance. We have 650 licensed women pilots in this country. Most of them would be of little use today, but most of them could be of great use a few months hence if properly trained and organized. And if they had some official standing or patriotic objective (rather than just to fly around an airport occasionally for fun) there would be thousands of more women pilots than there are now."[7]

Cochran understood the limits of 1940s-era gender equality and did not overstep the boundaries. Selfishly, she believed in advancement for herself but not for women in general. In her letter to Mrs. Roosevelt, Cochran stressed that she was not in favor of using women in combat. Women pilots, whom she referred to as "lady birds," should only be used in "back of the lines work." "Should there be a call to arms," she continued, "it is not my thought that women pilots will go out and engage in combat, for I'm sure they won't, but every trained male pilot will be needed in active service."[8] Russia, Germany, and England, she noted, were currently training women pilots for service. The United States should not wait to train its women.[9]

What Eleanor Roosevelt thought of Cochran's proposal is not clear, but certainly neither Franklin Roosevelt nor his administration acted on it. With the United States officially neutral and isolationism still a powerful belief, it is scarcely possible to see how Roosevelt could have pushed ahead with training women to fly military aircraft. Cochran, however, did not give up and she continued to speak out about her proposal. "If we became involved in war," she told the *New York Times* in May of 1940, "women may pilot mail and passenger planes at home and ambulance and troop transports abroad, but men, because of their advantage of steadiness and strength, will continue to bear the burden of combat flying."[10] As always, Cochran incorporated gender stereotypes in her thinking. Women would be used solely in an auxiliary capacity. As she later explained, "Women, being geared higher emotionally than men are not fitted for the strength required and the sustained strain involved in air fighting.... They would not and should not have to fight actively, but they would turn out men pilots well equipped to handle the fighting work in the air."[11]

Frustrated in her efforts to create an American counterpart to the British ATA, Cochran decided to join the war effort by flying for England. As was so often the case with Cochran, her decision was typically impromptu. At a luncheon for the members of the 1940 Collier Trophy Committee, Cochran sat with Army Air Corps Chief Henry "Hap" Arnold, and Clayton Knight, who was recruiting American pilots for the British Ferry Command. The Ferry Command handled all overseas ferrying and had established a base in Montreal in 1940. The organization was responsible for transporting American-made bombers, like the Lockheed Hudson, to England.

While discussing England's need for airplanes and pilots, Cochran volunteered to help Knight with his recruiting efforts. Hap Arnold suggested something bolder: why not fly a bomber for the British Ferry Command?[12] The idea thrilled Cochran. No American woman had ever piloted a Hudson Bomber. This would be another first. She would boost her own fame while also aiding the fight against Germany. Her flight would publicize the need for pilots, as well as deliver an airplane. By sending a woman into a combat zone, the British Ferry Command could make a powerful statement on its desperate need for pilots.

When it was proposed that Cochran fly to England, the idea faced both opposition and support because of her gender. The Ferry Command Officials in Montreal found her qualified, but remained skeptical because she was a woman.[13] Indeed, Cochran's critics and supporters could be separated into pro-woman and anti-woman factions. Supporters like Hap Arnold believed a woman should fly to Britain to dramatize and publicize the need for pilots. Critics including the Montreal Ferry Command leaders and various male pilots believed that a woman should not fly military aircraft because the Germans would shoot her down on sight or, even worse, it would "belittle" male pilots.[14] Cochran, however, had faced these prejudices before and had come to expect skepticism. As in the past, Cochran's gender was being used for an advantage in aviation. Women were still being used to exemplify the safety and ease of flying. Now, Cochran was expected to represent the brave American noncombatant ready to give England a hand. Indeed, Cochran's flight to Britain would be excellent propaganda.

Once again, Cochran's husband, Floyd Odlum, played a role in her adventure. Cochran and Odlum spent the first half of 1941 preparing for her flight across the Atlantic. Her first step was to obtain permission from the British government, with whom Odlum had connections. Cochran started her quest in Washington, where she contacted British authorities. They arranged an appointment with the Ferry officials in Montreal, Canada. According to Cochran, "They looked over my pilot's log book and flying record and said they could offer no objections as to qualifications but were frankly skeptical because of my sex."[15] The officials advised her to wait until they contacted her. "Time passed," Cochran recalled, "and no word came."[16] In the meantime, a change in British government favored Cochran's cause. Lord Beaverbrook, with whom Cochran and Odlum had socialized before the war, became the head of Procurement in England, and Odlum appealed to him to allow Cochran a chance to fly the Atlantic.

According to Cochran, Floyd had reservations about the flight. No doubt he feared for his wife's safety as she traveled over the ocean in the middle of a war. Despite his feelings, Odlum contacted Lord Beaverbrook and, through his friendship, Washington agreed to give Cochran a chance.[17]

After months of waiting, the British Ferry Command finally contacted Cochran and agreed to give her a test flight. Realizing the test would make or break her chance to fly across the Atlantic, she prepared by taking a captain's course with Northeast Airlines. As expected, Ferry officials put her through three days of difficult flying, during which resentful male pilots taunted her. When it was over, the Ferry officials gave Cochran an excellent rating, but with one exception. The test pilot reported that Cochran lacked the physical strength to handle the airplane's handbrake. Disturbed over what she saw as nitpicking, Cochran became furious when the Ferry officials agreed with his assessment.[18] In the end, however, Cochran agreed to a demeaning compromise. Captain Cochran would fly the Hudson bomber over the Atlantic Ocean, with navigator Stanford Grafton Carlisle making all the take-offs and landings.[19] Cochran herself thought the arrangement unnecessary and "unsatisfactory," but acquiesced simply to end the "battle of the sexes."[20]

Even with authorization, she still encountered male resentment. Male pilots tried to sabotage her take-off by breaking windows and removing equipment (including the life raft and the wrench for the oxygen system). Those aggravations did not stop her, and on June 20, 1941, she and her navigator left Montreal for England. The trip, it turned out, was uneventful except for a brief shower of tracer fire. She told the British press that her flight had been "a wonderful journey without incident," but later confided that "flying over the ocean was rather boring."[21]

In London, Cochran's flight produced mixed reactions. The press lauded her achievement and showered her with attention as if she were a movie star. As the *London Times* reported, "An American Lockheed Hudson reconnaissance-bomber has been delivered by a woman.... She is the first woman ever to have piloted a bomber aircraft across the Atlantic. Her achievement is remarkable when it is recalled that only a few years ago it was regarded as a great feat of endurance for anyone man or woman to make the crossing non-stop."[22] But ATA pilot Lettice Curtis found Cochran's visit counterproductive. "Jackie had entirely misjudged the wartime mood of the British people. [We were, however] grateful to her for her part in introducing us to a number of American women."[23] Meanwhile, some male pilots found Cochran's flight threatening. A week before Cochran landed, an American named A.R. Glancy had arrived in Britain as a passenger on an English bomber. Glancy, of the O.P.M. (most likely the Office of Preparedness Management), had been summoned by W. Averell Harriman, Minister to England. According to Glancy, Cochran's flight "completely defeated me as a hero and I am told created considerable disturbance among the pilots because, like me, they felt that if a woman could fly one of these bombers across there was no reason for them to boost themselves in the eyes of their friends."[24]

While in England, Cochran continued to exhibit her stereotypical attitude about women. For example, she played up her femininity to assure America that women could be involved in war without losing their womanhood. The English press, charmed by the glamorous Cochran, took notice when she explained she did not want to be photographed in wrinkled slacks. "I may fly bombers," Cochran said, "but I'm still feminine."[25] According to one London reporter, "The difference between Miss Cochran and the women of the uniformed services here was that she wore slacks and a jacket in which she did not want to be photographed."[26] Later, she posed for pictures in a "cool print dress worn with sheer silk stockings."[27] The article also claimed that Cochran arrived with three dozen silk stockings, a report she later denied.[28]

During Cochran's stay, she aided the British people. Observing the shortages in Britain reinforced her effort to help in the war. For example, Cochran advised President Roosevelt to send food supplies in great quantities. She also suggested that Americans sending packages to British friends should include animal or vegetable fats. Cochran regretted that she was unable to bring more food with her. During her stay, she took great pleasure in passing out oranges, cigarettes, and other unused items from the flight to the English people.[29]

Cochran remained in Great Britain for two-and-a-half months, traveling across the country assessing the bomb damage and studying the war effort. She made five radio broadcasts to Germany and, on one occasion, used the opportunity to boast of beating German pilots in a recent race. "I took pleasure," Cochran wrote, "in recounting that I took the 2,000-kilometer closed course air speed record away from the German army last year in an American army pursuit ship that was already being outmoded by later models."[30] Cochran also gathered information to take back to Washington.

Despite the lack of food and the less-than-favorable conditions,

Cochran seemed to enjoy viewing the bombing damage firsthand. She marveled at the huge bomb crater outside her hotel window and took home a piece of the parachute that had lowered the bomb to the ground.[31] By the time Cochran arrived in England, the bombing had subsided. Air raids, however, remained common. "While I was there," Cochran recounted, "we had some [bombing] 'alerts' and each time, if at night, I would hurry into my clothes and get to the roof of the hotel. I wanted to see whatever was going to come off."[32] There, she listened to the crack of the anti-aircraft artillery and watched the enemy airplanes turn back from their targets. Cochran's heart went out to the British people. She was saddened by the gutted houses and the thought of hungry British children. Yet she viewed the British people's plight with pride rather than pity. Cochran noted the "lines of care and determination" etched in the faces of her British friends and was impressed by their all-out defense against the enemy.[33]

Cochran was especially interested in Britain's women pilots whose wartime activities had brought her to England. While in London, she met with Captain Pauline Gower who headed the Air Transport Auxiliary. She had known Gower since the 1934 MacRobertson London-to-Australia Race. "It was no great prescience that originally made me urge the inclusion of women in the war," commented Gower. "It was more a desire to help in the way, I, and a number of women I knew, were best fitted to help."[34] By the time Cochran arrived, the organization consisted of 50 women and several older or disabled male pilots. When Cochran first inquired what ATA stood for, a ferry pilot replied, "Ancient and Tattered Airmen."[35] During her stay, Cochran spoke with most of the ATA girls and was impressed by their hard work, low accident record and cheerful attitudes. The ATA, however, was not without tragedies. In January 1941, Amy Mollison, Britain's only famous woman pilot and contender in the MacRobertson against Cochran, drowned when she bailed out over water.[36] Cochran remained in England throughout July, a month in which the ATA moved 4,718 airplanes.[37] Cochran returned home with a glowing report of the ATA's activities. "What I saw," she later wrote, "convinced me that there's a real job for every woman flier — British or American."[38]

Cochran returned to Canada in a B-24 bomber with fourteen other

5. Women in War

pilots: she sat on a floor made of loose boards and pallets. The 14-hour flight was rocky because heavy headwinds forced the pilot to fly low, close to the waves. After landing at Gander, Canada, an exhausted Cochran set out for New York City. Upon arriving, she said facetiously that she was going to sleep until noon the next day and did not want to be disturbed — not even by the president. At nine A.M. the next morning, however, she received a message. President Roosevelt had requested her presence for lunch.[39]

Just 48 hours after leaving war-torn London, Cochran sat down to lunch at the Roosevelt home in Hyde Park, New York. The luncheon was attended by Mr. and Mrs. Roosevelt, the president's mother, Sara, and Princess Martha of Norway. After lunch, Cochran followed the president into his office to discuss conditions in England. She particularly emphasized England's need for food. A few days later, Eleanor Roosevelt invited Cochran to the White House, where they discussed what service American women pilots could be to America. The Roosevelts' interest boosted Cochran's plans for organizing American women pilots for war service.

The Roosevelts' support led Cochran to meetings with other officials. On July 3, Mrs. Roosevelt wrote a letter to presidential military aid General Edwin Watson stating that "on the President's suggestion, [Jacqueline Cochran] will get in touch with you to make appointments for her with the Assistant Secretary of War and General Arnold of the Air Force. She wants to talk over with them the organization of women pilots of this country as they are organized in England."[40] Cochran met with Assistant Secretary of War for Air Robert Lovett, who sent her to Hap Arnold, who sent her to General Robert Olds. After running from one official to another, Cochran thought she had found the right man. General Olds was in the process of forming the American Ferry Command, which later became the Air Transport Command, and he "was agreeable to taking a few well trained women pilots."[41] But Cochran and Olds soon parted company over the issue of how the women would be organized. Olds wanted to follow the British example and incorporate a few experienced women pilots into the Ferry Command, while Cochran wanted a separate Women's Auxiliary Air Corps.[42] Cochran refused to cooperate with Olds, and she went back to General Arnold. According to Cochran, Arnold "agreed with me completely," but he advised her to put all plans on hold. "The time," Arnold said, "had not yet arrived to use women pilots at all." However, he added, "I should actively think about it for the present."[43]

Instead, Arnold suggested that she organize a group of American women pilots to fly for the British Air Transport Auxiliary.[44] As Cochran later explained to the commanding general of the American Flying Training Command, Major General Barton K. Yount, "At the end of October or the first of November, General Arnold telephoned me and asked me to go to England to fly for the Air Transport Auxiliary (ATA). After a conversation with Air Marshall [Arthur T.] Harris, who was then the Director General of the British Air Commission, it was decided that I undertake this work."[45]

Cochran agreed to recruit a small group of American women to ferry airplanes for the British ATA. "I was thrilled," she explained in a radio broadcast, "and very proud to think I had been chosen to organize the American girls."[46] Cochran would also return to Great Britain as the contingency's leader.

Almost as soon as Cochran agreed to join the British cause, the United States was attacked at Pearl Harbor. On December 7, 1941, the Japanese successfully surprised America's forces in Hawaii, sinking or damaging several ships and destroying American aircraft. The next day, President Roosevelt asked Congress for a declaration of war. Cochran wanted to organize women pilots for the U.S. Army but she felt obligated to carry out her plans for England. British officials, agreed, however, if the United States needed her, she would be free to leave.[47]

In January 1942, Cochran began recruiting American women for the British ATA. The American fliers would consist of 25 women who would serve for 18 months. While the American volunteers would be under the control and administration of the British ATA, they would be designated an American Unit, wearing their own uniform. The volunteers had to pass a medical examination and a flight test given by the RAF Ferry Command in Montreal, and then complete full training and conversion courses at ATA Headquarters in Great Britain.[48]

On January 23, Cochran sent telegrams to virtually every licensed women pilot. She explained the conditions of service and requested that anyone interested should contact her at her Fifth Avenue office.[49] The outcome surprised even Cochran. "I ... made an appeal for American girl pilots," she explained during a radio broadcast, "and the terrific response was far beyond my expectations."[50] Cochran, aided by her personal secretary and pilot Mary Nicholson, had the difficult job of choosing the 25 most qualified. She began with a personal interview. In the end, she chose women from 20 different states. Among those picked were Helen Richey, a co-pilot with Capital Airlines from McKeesport, Pennsylvania; 21-year-old Anne Wood, a flight instructor from Maine; and Emily Chapin, a 25-year-old file supervisor at the Standard Oil Company in New York.[51] Most of the women were in their early and mid-twenties, but some, like Richey, were in their thirties. The women trained for three weeks before sailing to Great Britain.

In England, Cochran quickly established that she was in charge of the American volunteers. When the British officials tried to insist on giving the women additional medical and written examinations, Cochran balked and used her influence to lighten the requirements. According to ATA member Anne Wood, "She [Cochran] always saw herself as our protector. She'd moan and groan and spend hours fighting our battles, battles that weren't really so important in the long run."[52] Petty, outspoken and confident of her own agenda, Cochran may have gone overboard on some issues. According to British ATA pilot Lettice Curtis, Cochran's "friends in high places and her U.S. aspirations made her altogether too high-powered for ATA, an organization which was not really big enough for her."[53] Cochran's tremendous ego made her difficult to work with, but she was also eager to make America proud of her women pilots.

After flight tests and a period of indoctrination, the Americans began working with the British ATA. The women made approximately five flights a day and flew 121 different types of aircraft, including the British Hurricanes and Spitfires. As an ATA pilot, Helen Richey became the first American woman to fly a Hurricane.[54] The women ferried airplanes from the factories to the bases, took airplanes to have their guns tested and flew partly damaged planes to repair shops.[55] At night, they returned to their hotel rooms or to homes of the London families who provided temporary housing for the American pilots. When the bombs were not falling, the women sampled London's wartime nightlife. At least three of the American women met their husbands during their services with the ATA, two of whom were widows before the war ended.[56]

The life of an American ATA member could best be described as "unpredictable."[57] When assigned to ferry an airplane, she never knew what kind of aircraft she would fly or how many flights she would make in one day. The women seemed to enjoy flying a variety of airplanes and aiding the war effort. In a letter back home, ATA flier Emily Chapman wrote: "This is really more like it, lots of work, tired at night, but a feeling of a little accomplishment."[58]

While serving with the ATA, American women experienced some hardships and initial resentment. They endured the food shortages and occasionally slept in unheated rooms with skimpy blankets. Some Britons resented the women because they were consuming food needed for the British. After all, the British people needed every available resource. To help ease these fears, the American women turned over their ration books to their landladies.[59] According to Lettice Curtis, the American women "were, for the most part, happy to take the ATA as it was — to accept the rather uncomfortable eating, heating and general conditions of wartime living in Britain and to become just ordinary ferry pilots. Amongst them were some of the best women pilots in the ATA."[60]

The American women experienced little, if any, discrimination. According to Emily Chapin: "The ATA is unique because men and women are exactly on the same basis."[61] British women were already ATA fliers, a factor which eased the acceptance of the Americans. In need of every available pilot, the British readily accepted women in their ranks. Still, the women were conscious that their gender could be used against them. "It's two strikes out of three against women, anyhow," commented American ATA Mary Hooper, "so we are more conscientious and more determined not to have anybody say 'Just another woman pilot.'"[62]

Despite their contributions, the ATA's American women did not receive much recognition. The *London Times* did not spotlight their activities or even mention their arrival. Overall, the work of the ATA was viewed as "unspectacular." According to the *Times*, the ATA "does not often attract the spotlight of public interest," which perhaps explains the lack of material on America's women pilots.[63]

The women were featured in at least one edition of *Fox Movietone*

News, in which Cochran praised their work. "I have seen the most magnificent thing I ever hope to see," Cochran commented in a feminine, lilting voice, "a very large group of women and they are doing a job, a most dignified work, I've ever seen done by women."[64]

Service with the British ATA provided American women pilots with a rare opportunity. When even very few American men had flown military aircraft, Jacqueline Cochran was the only American woman to have done so before 1942. The ferrying may have been unspectacular, but it was a "first" for American women. In England, the Americans provided a much-needed service for the Royal Air Force, releasing men for aerial combat. By September 1942, the ATA had flown 30,000,000 miles and delivered 100,000 aircraft.[65] Foggy conditions and damaged aircraft made the job of ferrying airplanes a dangerous endeavor. One of the Americans was killed. The lone death was Mary Nicholson, Cochran's secretary for the previous five years, who was flying a British Miles M. 19 Master II when the propeller suddenly flew off, causing her to crash.

Portrait of Jacqueline Cochran, wearing a white long-sleeved V-neck blouse and dark skirt, in October 1942 (photograph courtesy of Smithsonian Air and Space Museum, SI-78-15317).

Jacqueline Cochran spent six months in England, supervising the American volunteers. As flight captain, she set up a base at White Waltham and rented a flat in London. "She had little time for flying," commented Lettice Curtis, "our main memories of her are of someone who lived at the Savoy Hotel, wore a lush fur coat and arrived at White

Waltham in a Rolls-Royce, both noticeable because by [then] we had clothes as well as petrol and food rationing."[66]

According to Cochran, "I spent more time on the ground fighting administrative battles than I did in piloting planes in the air."[67] Acquainted with the London social set, Cochran was also a frequent guest at dinner and cocktail parties. Still, she worried about the food situation and stored grits and canned hams in a bank vault for protection against bomb attacks.

During her stay, two events stood out in Cochran's mind. She vividly remembered a deadly bombing raid that occurred shortly before she left. The screeching sirens forced her from her bed and outside into the night air. She wrapped a blanket around herself and sat on the front steps, watching houses explode and the window panes fall from her building. Cochran clung to her superstitious belief that her "time would come only when [her] number was up and went on [her] way without change of routine, except as required by conditions about."[68]

Cochran also recalled how she witnessed one of the most important decisions of the war involving air power. At a dinner party for Hap Arnold at the Savoy Hotel, Cochran watched as Arnold conferred with Air Marshall Arthur Harris, commander in chief of the Bomber Command of the RAF, and Admiral John H. Towers. Arnold approached her and asked if they could use her apartment for a private discussion. They invited Cochran to stay as they debated the question of night-versus-daylight bombing. In the end, the three men decided that the American force would bomb during the day and the British RAF would continue their nighttime raids.[69]

Before Cochran left England, she was summoned by General Arnold to help the 8th Air Force. Early in 1942, the combined chief of staff decided to create an American Bomber Force in Britain, designated the 8th Air Force.[70] When the "Eighth" established a Command Ferry Service in London, they turned to Cochran for her aviation expertise. Citing her "many years of flying experiences ... [and] experiences in this theatre," the commanding general of the American Air Forces requested that Cochran help formulate organizational plans.[71] As a commissioned officer (temporary grade) with the United States 8th Air Force, she devised plans for the 8th AAF Ferrying.[72] To Cochran, "It was a most interesting experience and took me into every kind of problem, including the laying out of airfields and the working out of tables of attribution for fighter planes."[73]

* * *

Since the war in Europe first began, Cochran had longed to join the war effort. She found her opportunity aiding the British. Long before the United States entered the war, Cochran delivered a bomber to the RAF, studied the use of women pilots on the home front and had organized American women flyers for the British. By late 1941, Cochran had completed her work with the British and now looked to America.

Cochran returned to the United States with a great sense of accomplishment. She had brought American women into the war effort and proved their usefulness. When Cochran left London in September, some of the American women felt a sense of abandonment.[74] According to Anne Wood, some women had criticized Cochran for doing very little ferrying work and leaving so soon.[75] Wood, however, stood up for Cochran's performance. "[Cochran] wasn't there in England for those eleven months simply to ferry airplanes from point A to B. She went to England to get us there, to prove her point, and to return to the United States."[76] Rather than abandoning the cause, Cochran sought to expand the program. She had her own agenda and was returning to the states to organize additional ferry pilots. The American ATA girls did not realize that Cochran was paving a way for their service with the American Army Air Corps.

6

Jacqueline Cochran and the WASPs

When the United States entered World War II, its military service and its industry faced the need to find and train workers and servicemen to meet the unprecedented demands of mobilizing the American economy and fighting a two-front war. One way of meeting these demands was to hire, enlist, and train women for jobs once deemed "men's work." Millions of women left home for work in factories, hospitals, and military bases. Another 350,000 either enlisted in the Army or Navy, or joined a military auxiliary branch like the Women's Air Service Pilots (a.k.a. the WASPs). These women delivered airplanes, helped train gunnery recruits, taxied officials, and tested aircraft. They put on men's flying coveralls, strapped on parachutes, and learned to fly the Army way. In 1942, over 25,000 women applied to the new women's pilot training program, 1,830 were accepted and over 1,100 graduated.

These women were able to fly for the United States because of Jacqueline Cochran. Cochran was a woman of action who foresaw America's entrance into the war. And when America joined the fight, Cochran wanted to be there. She also wanted a role for America's women pilots. For two years she had campaigned to include women fliers in the war effort. With the United States not yet at war, the Roosevelt Administration would not take this radical step, but after Pearl Harbor, the war crisis created an urgent need for trained pilots. Cochran soon found, however, that her ambitious plan to create an American version of the British ATA faced two obstacles. One was the appearance of competitors who hoped to organize a much smaller women's pilot program. The second was gender discrimination and male hostility within the Army Air Corps. To her dismay, the military did not believe that women possessed the strength, stamina or emotional competence to fly military aircraft.

6. Jacqueline Cochran and the WASPs

True to her nature, Cochran was determined to engage women pilots in military service. She had spent her life pursuing careers in business and flying. Now she would pursue war service with even greater skill and commitment. Her ambitious, overbearing personality both aided and hindered her cause. She often ignored military protocol and offended those in charge. But without her dogged persistence, women pilots would not have been utilized to the extent they were. Eventually, Cochran transformed a small group of women pilots into a large organization that benefitted the war effort.

Cochran's efforts on behalf of women flyers maintained her lifelong pattern of both challenging and sustaining traditional roles for women. She strongly believed in the abilities of women pilots, but felt their role should be limited. To help women pilots gain acceptance in the military, she stressed their femininity and temporary role. "Women, being geared higher emotionally than men," commented Cochran prior to the war, "are not fitted for the strength required and the sustained strain involved for air fighting.... They would not and should not have to fight actively, but they would turn out men pilots well equipped to handle the fighting work in the air."[1] By emphasizing that women would return to private life when the war was over, Cochran hindered the cause of sexual equality. By stressing the femininity of women pilots, she reinforced the notion that women were not suited to handle the jobs of male pilots in peacetime. She readily accepted the demeaning stereotypes popular in the media of women pilots as "feminine flyers" and "glamour girls in the sky." Still, had Cochran been more strident, there might never have been a WASP program.

Historians such as William Chafe and Sara Evans have argued that the World War II experience might have radically altered gender roles and sexual discrimination in American life. The war "marked a temporary retreat from the prevailing notions of women's capabilities and proper roles."[2] The crisis nature of the conflict, however, hindered this change in thought.[3] Because World War II was an extraordinary event, the break with traditional gender roles was easier to accept. After all, the war would end and so would the enlarged role of women in the work force and the military. In the end, the war did not produce any radical changes because the government, military, and the media insisted that women were to fill men's roles only for the duration of the war. Jacqueline Cochran shared these views, and she worked to ensure that, at the war's end, military flying would once again be for men only.

* * *

During World War II, American women worked as nurses, volunteered for the Red Cross, watched for enemy aircraft, and served as Army recruiters. They went to work in factories as machinists, welders and munitions makers, and riveters. In stark contrast to the 1930s, when women were discouraged from work, factory owners and government officials now implored women to join the war effort. This spirit of "woman power" gave women a chance to learn new skills and reinvent women's traditional roles. As William Chafe has explained, "The war made possible what no amount of feminist agitation could achieve: it propelled women into a new and wider sphere of activity."[4]

Women also joined the military. Until the start of the 20th century, the services employed women only as civilian employees who took on jobs such as nurses, laundresses, clerks, and emergency aides. In 1901, when Congress established the Army Nurse Corps, women were finally allowed to hold military rank and full membership in the armed forces. During World War I, the Army opposed proposals to bring more women into its ranks, although the Navy did enlist 13,000 women as clerks with full military rank and status. After the war ended, women did not serve in the armed forces.

Between the two world wars, the military debated the role of women in the armed services. The General Staff considered establishing a women's corps, but plans were shelved until October 1939, a month after Germany invaded Poland.[5] Even then, the military failed to produce a workable plan. Two years later, Congresswoman Edith Rogers introduced a bill to establish the Women's Auxiliary Army Corps or WAAC. The bill stalled until the Japanese bombed Pearl Harbor. It was finally passed in May 1942, three months after Chief of Staff George Marshall chose Oveta Culp Hobby as WAAC Director. In 1943, the WAAC was integrated into the Army and renamed the Women's Army Corps or WAC.

Unlike the Army, the Navy accepted women on the same basis as male Navy reservists.[6] By late July 1942, women could join the WAVES

6. Jacqueline Cochran and the WASPs

(Women Accepted for Volunteer Emergency Service) or the SPARS (for *Semper Paratus*) under the Coast Guard. Women also aided the Marines, although without an acronym.[7]

During World War II, American women were anxious to join the war effort. Caught up in the patriotic fervor, women joined the services in unprecedented numbers. By 1945, 350,000 women would serve either in military organization such as the WACs or in auxiliary organization such as the WASP. The historian D'Ann Campbell has called "making women soldiers ... the most dramatic break with the traditional sex roles that occurred in the twentieth century."[8] Women not only joined the Army and Navy in record numbers, but they were able to perform more meaningful work than at any time in the past. For the most part, they continued to serve in supporting roles to men, and were subject to demeaning regulations. Usually, too, their work was "suddenly endowed with femininity and glamour for the duration."[9] Nevertheless, the battlefield needs of North Africa, Europe, and the Pacific made it possible for women to take over "men's work." One example of this willingness to employ females was the use of women as pilots.

The decision to use women pilots was prompted by two important factors. First, there was a severe shortage of pilots to fly combat missions and serve in supporting roles, such as ferrying or training. Secondly, the Army Air Corps looked favorably on the past role of women in aviation. In the 1930s, as Susan Hartmann has noted, women "had been highly visible as the commercial aviation industry employed them to demonstrate and sell planes and sponsored them in races."[10] An additional factor was Jacqueline Cochran. Her persistence and willingness to work within the system helped secure a place for women pilots.

Cochran had grandiose ideas. While she admired the work of the British ATA and even used it as her model, she sought bigger things for America. Cochran wanted to build a large organization of women pilots and train them to fly various types of aircraft. She also wanted to expand their duties beyond ferrying. Despite Cochran's willingness to embrace gender stereotypes, she believed that women pilots could fly just as well, if not better, than their male counterparts.

Cochran knew she faced many obstacles. Many Air Corps officers saw no need for women pilots and were skeptical of their abilities.

Jacqueline Cochran

Cochran, however, had the support of General Hap Arnold, the head of the Air Corps. Before Cochran had left for England, he had assured her that if America decided to use women pilots, Cochran would be in charge of organizing them. Cochran's contract with the British had been drawn up with this possibility in mind. Only a few months after the bombing of Pearl Harbor, General Arnold called Cochran back to the United States to organize women pilots.

Cochran soon found that she faced another challenge. While she was in England, Nancy Love, a long-time rival, had been chosen to head the Women's Air Force Ferry Service (or WAF). The 28-year-old Love was a veteran pilot who possessed 1,200 hours in flying time and had ferried airplanes to the Canadian border for delivery to Great Britain. In May 1940, she had written to the Army Air Corps Plan Division offering to recruit women pilots for the Ferry Command.[11] As the wife of Robert Love, the administrative assistant to the chief of staff of the Ferry Command, she undoubtedly enjoyed important connections and the support of the Ferry Command.[12]

When Cochran learned of the WAFs upon returning to the United States in September 1942, she was furious. She felt cheated and betrayed and even wondered if she had been tricked into staying in London three days longer to aid Love's plans.[13] According to Cochran, she had been ready to leave England when she received word that General A.H. Frank of the 8th Air Force wanted to see her. Three days later, she attended the General's dinner where he thanked her for helping establish the 8th Air Force. Now, after her return to America, she viewed the dinner as part of a conspiracy to delay her return and prevent her from discovering Love's appointment.[14]

Cochran was undoubtedly angry at the possibility of being pushed aside by Love, but she also objected to Love's plans for women pilots.[15] Love envisioned a small group of elite women pilots who would be used in a very limited capacity. As head of the WAFs, she chose only a few, experienced women who would serve the Ferry Command as civilians. Love did, however, favor granting women pilots' military status under the Women's Army Corps. In other words, women pilots would be integrated with the Air Corps , which was still part of the Army. This suggestion seemed like a logical idea and it was endorsed by General Harold

George of the Air Transport Command (ATC) and Colonel William Tunner and General Robert Olds of the Ferrying Division.[16]

While Love was content to limit the WAFs to 25 women ferrying military aircraft, Jacqueline Cochran was thinking on a much larger scale. She envisioned training several hundred women pilots under military supervision to assume a wide array of flying missions. Initially, Cochran also wanted women pilots to be granted military status and to be separate from the Women's Army Corps. The controversy over militarizing women within the WACs had caused Cochran to break relations with General Olds before her trip to England.[17] This conflict may have prompted Olds to side with Love over Cochran.

As always in her various careers, Cochran went to the top with her project. She conferred with General Arnold, who was "mad all over" about the creation of the WAF.[18] "As he said in no uncertain terms, [they had] gone around his desk to the Secretary of War with a project he hadn't approved except in principle and was not going to approve in detail."[19] Later, in a 1945 interview, General Arnold admitted to seeing the September 3 memorandum recommending the use of women to ferry planes.[20] According to Arnold, "I had instructed that the plans be prepared for use of women pilots but contrary to my intentions these plans were activated without first being brought back to me personally."[21] Clearly he wanted Cochran in charge of a woman's unit. Early in 1942, for instance, Arnold had written to Major General John F. Curry praising Cochran's success in England and advising him to formulate a role for her when she returned to the states.[22] Arnold, no doubt, viewed her celebrity status as an advantage for the war effort, but he also appreciated and admired her ability as a pilot.

In letters to Arnold—some marked urgent and important—Cochran argued against making women pilots part of the WAC, calling it "bad ... from the organizational standpoint, and contrary" to what Arnold had stated in the past.[23] "In addition, it would wash me out of the supervision of the women flyers here rather than the contrary, as we contemplated."[24] If women pilots were incorporated into the WACs, Col. Oveta Hobby, not Cochran, would be supervising women pilots. Rather, Cochran argued, Love's ferry pilot organization was "just one segment of a larger job to be done."[25] She envisioned a new organization including training facilities and a larger role for women. She also proposed a powerful position for herself as the program's director. Cochran wanted to train women pilots to serve with American forces, both on the home front and in England. She assured Arnold that with the proper authority and facilities she could provide over 1,000 well-trained women pilots. Within three months, she hoped to have 1,000 women capable of ferrying small aircraft and at least 100 women capable of ferrying fighter airplanes.[26] She urged the general to "announce immediately that a broad program for use of women pilots has been under discussion between the two of us for the past year [and] that you have asked me (rather than Nancy Love) to take charge of such broader program under your control and direction."[27]

Cochran readily admitted to Arnold that the clash with Love was part of a personal rivalry. According to Cochran, she told General Arnold that Love had harbored a "deep hatred" of her for a decade.[28] Supposedly, Love resented Cochran because she "had done things [Love] wanted to do."[29] For example, when Love's plans to fly a B-17 to Europe fell through, Cochran said she was blamed. After this incident, Cochran wrote Arnold, "Her [Nancy Love's] hatred grew to the explosion point where she lost her sound reasoning powers and was against everything I tried to do."[30] Cochran's appeal to Arnold achieved immediate action.

Arnold called the officers of the Ferry Command to his office and, according to Cochran, told them "to rework the project to my satisfaction."[31] Now having the authority to reorganize the WAF program, she outlined her ideas in a memorandum to General Olds. Rather than dissolve the WAF, she proposed establishing a separate training school under Cochran's direction. To find applicants, Cochran wanted to survey available women pilots and interview the most qualified. She also requested that orders be drawn up directing her to select squad leaders, make arrangements for housing, establish rules and regulations and create a list of airplanes to be ferried.[32]

General Arnold responded by establishing the Women's Flying Training Program, a civilian organization under the 319th Army Air Forces Flying Training Detachment (AAFFTD). In a memorandum to General Barton Yount, Arnold ordered that Yount take "immediate and positive action to augment to the maximum possible extent the training of women pilots. The Air Force's objective is to provide at the earliest possible date a sufficient number of women pilots to replace men in every non-combatant flying duty in which it is feasible to employ women."[33] On September 14, 1942, he announced Cochran's appointment as director of the new program. The program was established to train women pilots to fly military aircraft. Upon graduation, the women would become part of the WAFs.

Cochran's appointment met with some opposition. Colonel William Tunner of the Ferry Division did not welcome additional women pilots. At the time, the Division did not have enough jobs for male pilots and did not need more women.[34] Tunner also questioned Cochran's motivations and qualifications, pointing out that she had had "no hand in the training procedures" of the ATA.[35] Cochran, however, had never intended to train women pilots. She saw herself as an administrator and an organizer, and, though qualified to instruct flyers, she would leave hands-on training to others.

Despite Tunner's opposition, Cochran began training women to join the Ferry Division. She was assigned to the Training Command, headed by General Barton Yount, and officially served as an assistant to Colonel Luke Smith, general staff officer for pilot training. Almost immediately, Cochran and Smith began to clash. Two weeks after Cochran arrived in Washington, Smith assigned her an assistant, Lillian Conner. Cochran found the young woman incompetent and promptly sent her back to Smith. The colonel then made Conner a liaison between himself and Cochran, even though, as Cochran pointed out, "[Smith] and I had adjoining offices."[36] A week later, Smith transferred Cochran out of his office. He sent her to the Training Command at Fort Worth, Texas. While Cochran suspected that her new appointment was some "sort of banishment," she welcomed the opportunity to work in the field.[37]

Cochran threw her energy into organizing women pilots. A successful business entrepreneur, she was at her best formulating plans and working out details. She had previously mapped out plans for the American ATA contingency and had helped establish the 8th Air Force in England. Following her transfer to Texas, she immediately scanned the area for a suitable base for her training program. Against the wishes of Colonel Smith, Cochran chose the Houston Airport. She began leafing through applications, searching for her first candidates. The training itself would be conducted by civilian instructors, contracted by the army.[38] Some of the school's instructors were furnished by Aviation Enterprising, Inc., a private flying school located at Houston airport.[39]

General Arnold directed Cochran to train 500 women, but later the number was increased to over a thousand. To qualify for the training program women had to be between the ages of 21 and 35, have a high-school education, be at least five feet tall and have 200 hours of flying time. Cochran "looked for clean-cut, stable young women who could show flying hours properly noted and certified in a log book."[40] While interest in the program was enormous, few women pilots possessed 200 hours of flying time. To compensate for this problem, Cochran lowered the standards.[41] In fact, she cut the number of flying hours by more than 80 percent, or from 200 hours to only 35. With the new standards, some 25,000 women applied to the program. Cochran hand-picked 29 women to form the first class of trainees. Cochran also struggled to find equipment and support. She later recalled, "We started with the claptrap equipment consisting of every conceivable sort of primary and basic training planes."[42] Eventually, she was able to acquire about 200 airplanes including trainers such as PT-17s, PT-19s, BT-13s, AT-6s, UC-78s, UC-43s and UC-18s.[43]

Meanwhile many men remained skeptical of women pilots. According to Geri Lamphere Nyman, a member of the first training class, "There was a memo out, saying get rid of these women as quickly as possible. We know they cannot fly military aircraft, but we must give them the opportunity."[44] Even General Arnold, a strong supporter of the program, had doubts that a woman could fly a heavy bomber in adverse conditions. In 1944, he told a group of WASPs, "I didn't know in 1941 whether a slip of a young girl could fight the controls of the B-17 in the heavy weather."[45]

Designated Class 43-1, the women first met with Cochran in Houston on November 15, 1942, at the Rice Hotel.[46] The women of Class 43-1 had come from several different states. Some were wives and mothers, with husbands in the service. Others were single and looking for a way to join the war effort. All of the women met the original qualifications. One, a flight instructor, possessed 762 hours of flying time. The average number of flying hours was 315.

6. Jacqueline Cochran and the WASPs

To many of these young recruits, Jacqueline Cochran was viewed as a glamorous celebrity. No doubt many of them had read about her successful air races, and some were familiar with her cosmetics company. One of the new trainees, Marjorie Kumler, was waiting with other women pilots amidst uniformed marines, infantrymen, WACs and Red Cross workers when Jacqueline Cochran arrived. According to Kumler, "There was a flurry at the door when [she] came in, followed by the flight contractors."[47] Cochran, she recalled, looked "neat and tailored, even to her lapel ornament."[48] (The ornament was a small silver pin shaped like an airplane propeller with a large rosette diamond at its center.)

Well dressed and confident, Cochran spoke to the trainees with business-like professionalism. She explained to the women pilots the importance of the program and assured them that their contributions to the war effort were needed. "You girls," Cochran explained, "are the first women to be selected for training by the Army Air Forces. You are all experienced pilots.... If things don't run smoothly at first, just remember that you will have the honor and distinction of being the first women to be trained by the Army Air Forces. You are badly needed, and I hope that you will be out of here in two and half to three months at the most."[49]

Afterwards, Kumler and the other women were issued flying suits, regulation wool sweaters and A-2 jackets. They then boarded a sky-blue bus destined for Howard Hughes Airport. At the airport, the group was met by Cochran's assistant, Chief Administrator Officer Leoti Deaton, a 40-year-old Red Cross administrator. Deaton emphasized the importance of their work and cautioned the women to keep their activities secret. After taking an oath, the women were dismissed and began combing Houston for housing.[50] Some of the women rented rooms in the local boarding house, while others found rooms in private homes. At least three of the more well-to-do women rented a suite in a luxury hotel.[51] In this first night in Houston, they anxiously awaited the next day's assignments.

The first day of training began on a cold, damp November morning at the Houston Airport. At 7:45 A.M., class 43-1 reported and listened to Leoti Deaton introduce their instructors. The chief pilot, a male instructor, then explained the first step of the program. He told the women that they would each "fly 20 hours, most with an instructor, and then ... get a check ride with an Air Forces pilot.... You may do ok," he cautioned, "or you may be sent home."[52] Only one woman failed this initial test.[53]

For the next 23 weeks, the women attended ground school and flew over 20 different types of airplanes. A typical day started at 6:30 A.M. The women would dress and then catch the bus to the field at 7:00. After a 7:30 breakfast, they would work on the flight line from 8:00 A.M. to 1:30 P.M. Following a 45-minute lunch, they attended ground school from 2:45 until 4:45, and ended the day with an hour-long calisthenics workout. At 6:30, the women ate supper, then boarded the bus for home.[54] They worked ten hours a day, six days a week mastering instrument flying and navigation. Although civilians, they lived like regular soldiers. The women drilled, marched to class and ate Army food.[55] Because of the newness and skepticism about the program, the military referred to class 43-1 as the "guinea pigs." Officially designated the Women's Flying Training Detachment (WFTD), the women referred to themselves as "Woofteddies." The nickname, a somewhat proactive acronym, probably did not aid their campaign to be taken seriously as competent pilots. Overall, the women seemed to enjoy their weeks in the 319th. They wrote patriotic songs and published a newsletter that included news, features and original cartoons. The women also welcomed the whimsical emblem drawn for them by Walt Disney. Known as the "Fifinella," the winged, cartooned female gremlin became a symbol, as well as a nickname, for the WFTD.[56]

During training, the women earned $150 a month and the satisfaction of helping the war effort. That said, the women did have complaints. They disliked the food, instructors who cursed, and the size 44 flying suits. Referred to as "zoot suits," these coveralls were especially uncomfortable for the five-feet-tall women.[57] Some of the women also disliked Deaton, whom they resented for "telling them every move to make."[58] No doubt Deaton had her hands full with this first class. Meanwhile, Jacqueline Cochran had returned to Fort Worth after the training command was established.

As head of the training school, Cochran established a staff and made arrangements for it. While the women trained, Cochran tested airplanes

6. Jacqueline Cochran and the WASPs

and served on the Air Force's board for training curriculum.[59] Deaton was left to handle the headaches of the program. Since the women were civilians, they were not subject to military orders of discipline. Deaton, who seemed to lack the tenacity of Cochran, had some difficulty dealing with the women's complaints and the testy attitudes of some of the instructors.[60]

By the end of the month, the Flying Training Detachment began to take shape and was noticed. Despite the skepticism, the first class had made sufficient progress to inspire another. According to G.C. "Brownie" Brown Kindig, a member of the first training class: "We were the guinea pigs. In fact, we were told if we made it there would be further classes, if we didn't make it and couldn't pass the tests and requirements necessary their whole program would fail. Needless to say we did our best and we succeeded."[61] In the second week of December, class 43-2, which included 60 women, reported to the Houston airfield. A new class reported each month thereafter.[62]

In Fort Worth, Jacqueline Cochran continued to work to better the program. Her first priorities were to find better equipment and a better location than the Houston Airport. When the British abandoned a training base at Sweetwater, Texas, Cochran "seized upon it" for the women and the second class of trainees were transferred by air to the permanent home.[63]

Cochran's other prior-

Official portrait of Jacqueline Cochran in her Women's Air Service Pilot's Uniform (photograph courtesy of Smithsonian Air and Space Museum, SI-2005-29843).

ity was to keep her organization separate from the Women's Army Corps. In December 1942, the military considered integrating women pilots into the WAC. The idea infuriated Cochran, who did not want pilots under an Army commander. Cochran believed that the women who joined her training command were somehow better than women who joined the WACs. As she told WAC commander Colonel Oveta Hobby, "They're a different breed of cat" and she "couldn't see throwing [her] girls in with that bunch."[64] And Cochran, true to her ambitious nature, did not want to lose her position. If women pilots were integrated into the Women's Army Corps, Hobby, not Cochran, would be in charge. An adamant Cochran declared, "It was my job and I wanted to keep it."[65] Cochran defended her position in a letter to Brigadier General Kenneth McNaughton, a member of General Arnold's Air Staff. In a well-crafted argument, Cochran explained that "putting the women pilots in the WAACs would be like putting the Air Corps in the Infantry."[66] Cochran was especially disgusted that someone without aviation experience might take charge of the women pilot's program. "It certainly wouldn't seem advisable to turn a branch of the Army Air Forces over to someone who admittedly knows nothing at all about the subject of aeronautics, and then expect to obtain good results."[67] The idea was dropped for a time, but it resurfaced again in 1943.

Despite the training program's success, Cochran spoke little about her time as director. With headquarters in Fort Worth, she served in an administrative capacity. She made inspections and decisions about equipment, types of training, and duties for the pilots. On December 17, Cochran accompanied General Barton K. Yount to Houston to inspect the first class. Yount referred to it as "outstanding" and was impressed by Cochran's work.[68] Two months later, Cochran returned to the base to investigate morale problems, which mostly involved petty complaints about Deaton. Whatever the occasion, Cochran's appearance generated much excitement among the trainees. Famous, glamorous and dedicated to the cause of women pilots, Cochran was viewed as a heroine by "her girls." On one of these visits, trainee Margaret Boylan recalled how Cochran "came roaring in and we all went to meet her and gather around her airplane."[69]

In late April, Cochran made plans to attend the graduation of class

43-1. But when she suddenly became very ill, with what she termed "abdominal adhesions," she feared she might miss the event. Stomach problems had plagued her since childhood, when an inept surgeon botched an appendectomy. Shortly after class 43-1 completed training, Cochran checked into St. Vincent's Hospital in Los Angeles. She stubbornly refused an operation, even though her doctors warned her that the problem would probably recur. Instead, she returned to her California ranch to rest. As the day of graduation neared, she still could not sit upright, but she made the 200-mile journey to the Phoenix Airport in a rented hearse from the local funeral home.[70]

On April 24, Cochran stood proudly on the stage platform and watched class 43-1 march in a graduation procession.[71] By that time the third class had begun training. They, along with class 43-2, gathered to watch the ceremony. The graduation was also attended by several generals and members of the press. Byrd Granger, one of the trainees, recalled how the "reporters, cameramen, and newsreel crews scrambled for a better position, staying along the path of the marching girls to the point of near disaster."[72] The graduation was an important event for Cochran, who realized that a successful first class would help justify the program.

Of the 29 women who started with class 43-1, 23 completed the training. After graduation, these women became part of the Women's Air Force Ferry Service and were assigned to bases of their choice. Most chose bases close to their homes and families.[73] As part of the WASPs, the women took some additional training and began ferrying aircraft inside the United States. Because of America's romance with aviation, these women were often viewed as glamour girls in the sky. One unnamed ferry pilot demanded a more accurate portrayal from the press. "Don't make you story kittenish," she warned, "because our job isn't in the least.... We're ferry pilots first and women afterward."[74] Cornelia Fort, one of the original WAFs, also attempted to correct the glamour image. "We get up in the cold dark in order to get to the airport by daylight. We wear heavy cumbersome flying clothes and a thirty-pound parachute. Any pilot can tell you how glamorous it is."[75]

Nancy Love concurred, pointing out that the women had very little time to themselves. The women would ferry airplanes to Army bases, rest six or eight hours and fly again.[76] Since the beginning of the WAF program, Love had fought against portraying women flyers as glamour girls. According to one article, "There will be no fuss and feathers — no phony glamour about the WAFs as a far as Director Nancy Harkness Love is concerned. That immediately becomes clear when you talk to her."[77] Unlike Cochran, Love did not stress the temporary nature of women ferry pilots or the proper place of women.

Meanwhile, the training command continued to prepare more women to join the WAFs. By May all the "Woofteddies" were being trained at the new facility, Avenger Field in Sweetwater, Texas. By June 25, the WAFs had expanded to 158 members, and this number included the 25 women organized by Nancy Harkness Love, the 23 graduates from class 42-1, the 43 graduates from class 42-2, and the 67 graduates from class 42-3.[78]

As the WAFs grew, the organization began to experience problems stemming from its civilian status. Unlike other wartime organizations, such as the WACs or the WAVES, the WAFs were not a military unit. The women pilots did not have uniforms and were not subject to military orders or discipline. Disturbed by the appearance of the Women Ferry Pilots, the Air Transport Command suggested that the women adopt the WAC uniform. The officials even sent a staff photographer to snap pictures of Nancy Love wearing the uniform and sent them to the secretary of war.[79] This idea, apparently, never materialized. When the WAFs became part of the WASPs in 1943, General Arnold gave Cochran the responsibility of designing a uniform. Using her own fashion sense, Cochran had a uniform made from her own design and color choice, Santiago blue.[80]

* * *

In 1943, the women's pilot program was again reorganized in Cochran's favor.[81] Since 1941, General Arnold had envisioned a program with Jacqueline Cochran at the helm. He admired her flying skills and was impressed by her work with the British ATA. Arnold also understood the advantages of Cochran's name recognition and glamorous image.

Moreover, Cochran's willingness to adhere to traditional roles was a key to her success. She stressed that women pilots would not be used in combat and not in place of men, but to release male pilots for combat duty and overseas flying. She also advised the women not to challenge traditional gender roles. She suggested, for instance, that women pilots allow men to carry their parachutes, if the latter offered. "If you are going to run around trying to act like men, they are going to treat us like men. If we act like ladies, we'll be treated that way."[82]

With the success of Cochran's training program and the growing number of WAFs, General Arnold decided to merge the two programs.[83] On August 20, 1943, the War Department announced that all women pilots under the Army Air Forces would be known as WASPs or Women's Air Service Pilots. The Women's Auxiliary Ferrying Squadron (WAFS), under Love, and the Women's Auxiliary Flying Training Detachment (WFTD), under Cochran, were merged into a single organization. General Arnold named Cochran as director and Love became the executive for WASP ferrying operations for the Air Transport Command. While Cochran moved into a Pentagon office in Washington, D.C., Love was transferred to Cincinnati, Ohio. According to WAF pilot Adela Riek Scharr, "It was an unhappy day for Love."[84] Once the head of her own program, Nancy Love was now under Cochran's constraints.

Even after the reorganization, there was some question as to Cochran's authority. Referring to both women as "photogenic female flyer[s]," *Newsweek* pointed out that the Air Corps was divided over which women would outrank the other if women pilots were militarized. According to the Air Transport Command, "Cochran's job was merely advisory and not superior to Mrs. Love's executive post."[85] Meanwhile the Air Force maintained that Cochran had the "highest authority.... If the Air Transport Command is not already aware of this, they will have to be made aware of it."[86] Despite these reports, Nancy Love contended that there was never any battle between herself and Jacqueline Cochran.[87] To dispel the notion of rivalry, she pointed out the difference between their jobs. According to Love, "Miss Cochran's was an administrative job and mine was an operational one."[88]

While the WASP remained a civilian organization, Cochran (aided by General Arnold) had considerable authority. Three months later, the War Department sent out an official description of the WASP uniform, the Santiago blue jacket, skirt and beret designed by the director herself.[89]

Jacqueline Cochran

The first training class, officially designated as WASP, arrived at Avenger Field in September 1943. The class consisted of 112 women between the ages of 18 and 25. Some of these women, like Leona Golbinec, had been interviewed by Cochran for the ATA and had now been accepted into the WASP program. According to Golbinec, Cochran would not take her in 1942 because she was in college. "She wanted me to finish my degree first. She didn't have a degree, but she really thought that education was very important."[90] Cochran's lack of education hindered her since her youth. Remembering her own struggles, Cochran now viewed Miss Golbinec's educational opportunity as more important than her participation in the war effort.

In 1943, Leona Golbinec became a member of class 44-W-2, the second group of trainees scheduled to graduate in 1944. At Sweetwater, little had changed, except that the Woofteddies were now renamed the "WASPs." Most of class 44-W-2 were pleased with their instructors and excited about flying various types of aircraft. Although trainee Jean Hascall Cole was enthusiastic about the program, she confessed that some of the duties were dull and mundane. According to Cole, "Many hours were spent in the classroom, on the marching field, in daily calisthenics, and in mundane chores such as cleaning our barracks, washing clothes, or writing letters home."[91]

Nevertheless, the women found themselves with a unique opportunity to fly the military's top aircraft, an opportunity made possible by Jacqueline Cochran. While Cochran maintained traditional attitudes about women in the military, she challenged traditional roles when it came to women in the air. She maintained that her girls could fly any airplane that male pilots could handle. On the other hand, she continued some discriminatory practices by using women to shame men into action. When male pilots refused to fly the B-26, Cochran suggested that "putting a couple hundred girls on it would cure the men."[92]

The B-26 story exemplifies the considerable skills of Cochran and other women pilots. The B-26 was a troublesome airplane. The Army Air Forces referred to the medium bomber as "the most difficult airplane to fly."[93] When male pilots refused to fly the B-26, General Arnold asked Cochran to make a test flight. After flying the plane, she suggested that WASPS trainees be admitted into the B-26 program. Cochran knew that airlines had promoted airline safety by encouraging women to fly. Now she attempted to promote the safety of the B-26 by using the same approach. According to the Army Air Forces, when Cochran made the suggestion, not a single officer believed the women would succeed.[94]

Some 20 WASPS entered the B-26 program at Dodge City, Kansas. Halfway through the course, Cochran arrived at the base to inspect the training. Pleased with the program, she returned to Washington and anxiously awaited the results. Surprisingly, Cochran found the training facility congenial to the women pilots. According to the Army Air Forces, the Dodge City base "considered the women pilots good for morale and they were delighted to have them."[95] The women's opinions of the troublesome bomber were mixed. Anna Mae Petteys hated the B-26, and although disappointed when she washed out, she contended that "it would have been very unpleasant for me to continue to fly it."[96] Six other WASPs also failed the program. A total of 13 women, however, passed with excellent ratings. While the pilots found the training hard and challenging, some fell in love with the B-26.[97] When the training was completed, the women outscored the men in ground school. Afterward, the B-26 lost its frightening reputation. When the 13 "girl" graduates mastered the B-26, Cochran had proven her point and helped move along the military's training program.

* * *

As director of the WASPs, Cochran found herself the head of a growing organization with increasing responsibilities. While supervising training, she concentrated on one major goal, granting military status for the WASPs. Despite her initial concerns about organization, she had agreed to train the women as civilians. Then, when the women pilot program evolved into the WASPs, the women remained civilian employees. Not only did Cochran want the WASPs militarized, she wanted the change to be on her terms. The fight for militarizing women pilots was Cochran's greatest challenge as WASPs director.

7

The WASP Experience

We are the Yankee Doodle Pilots, Yankee Doodle Do or die; Real live nieces of our Uncle Sam, Born with a yearning to fly; Keep in step to all our classes, March to the flight line with our pals — Yankee Doodle came to Texas just to fly the PT's — We are those Yankee Doodle Gals!—WASP Song, World War II

In 1970, Jacqueline Cochran referred to the WASP program as one of "the most important accomplishments of [her] flying career."[1] Inspired by a mixture of patriotism and her desire for fame, she took the 25-member WAFs and turned it into an organization of over 1,000 service pilots. The women were assigned flying duties in conjunction with the Army Air Corps. They ferried a variety of planes including heavy bombers and speed pursuit planes, helped train gunnery recruits, transported military personnel, and helped train soldiers for combat. Cochran later reflected that "what we did back in 1943–44 was something that any women could have done, given the ambition and drive that we had, but it was unusual for the time."[2] To Cochran, the program was "the greatest opportunity ever offered women pilots anywhere in the world."[3]

The history of Jacqueline Cochran's WASPs is a remarkable story of courageous women who performed their duties under challenging conditions. It was dangerous work, and nearly 40 of the women died in crashes during the war. Always, too, the WASPs faced challenges because they were women in a male-dominated military and a male-dominated field of aviation. Male coworkers were skeptical, even hostile, and the press was condescending. As always, Jacqueline Cochran both advanced and hindered the cause of equality. She built the WASPs and was proud of its accomplishments and the expanded role for women in time of war:

7. The WASP Experience

yet, all the time, she believed that women's roles in warfare should be limited. The war emergency gave Cochran a chance to demonstrate her skill as an organizer and the ability of 1,000 women pilots. But this experience in no way altered her hostility to feminism and feminist ideas.

* * *

The visionary behind the WASP program was Jacqueline Cochran. As director, she built a large organization in which women pilots could aid the war effort. When the Army Air Corps first considered using women pilots, General Hap Arnold envisioned them only ferrying light trainer planes.[4] Cochran, however, expanded the types of aircraft women could fly and duties they could perform. After Cochran established the training base in Sweetwater, her primary objective was to train women to ferry many different types of aircraft in order to release as many men as possible for combat duty. The women ended up flying a variety of airplanes, including heavy bombers and pursuit planes. Through Cochran's influence, WASPs also assumed duties beyond ferrying. They helped train gunnery recruits, transported military personnel, instructed other women pilots and trained soldiers for battlefield attacks.

Jacqueline Cochran was a determined woman with a strong commitment to women pilots and her position as their leader. She was very possessive about "her program" and "her girls." An experienced businesswoman, Cochran used her organizational skills to set up and implement the program. Running the WASP program was not an easy task. Because of the program's civilian status, neither women pilots nor their director was subject to military orders. Cochran's stubborn, gutsy determination both helped and hindered the program. Through her persistence, she designed and expanded the role of women pilots.

As director of the WASPs, Jacqueline Cochran served in an administrative capacity. While she emphasized the importance and effectiveness of the women pilots, she spoke little about her own role. Her official duties included creating policies and procedures relating to planning, recruitment, training and assignment of all women pilots in the Army Air Forces.[5] She had begun recruiting women even before America entered the war. She gathered names of licensed female pilots and sent them questionnaires. Later, when the WASPs were formed, Cochran handled the correspondence relating to the program. After receiving a letter of interest from a woman pilot, she responded with a description of the program, qualifications and necessary forms. If a woman met the initial qualifications, Cochran set up a personal interview, and then either rejected the candidate or assigned her to a training base. This "final hurdle," the interview with Cochran or one of her representatives, was referred to as "one of the most rigid" prerequisites.[6] Cochran also made decisions regarding the training and utilization of WASPs.

While her feminine demeanor helped win acceptance of the program, Cochran was not without critics. When the WASP program was first established, for instance, she was criticized for taking three weeks to initiate the program. This seemed excessive, since she made the move from her offices in Houston to the Pentagon in a mere two days.[7] On the other hand, she needed time to work out the new duties of women pilots. Whether she could have accomplished more in less time is difficult to prove. Considering what Cochran did accomplish, the criticism is trivial. More serious complaints came from the Ferry Division.

At times, Cochran's relationship with the Ferry Division was stormy. She caused some resentment when she reorganized Nancy Love's WAFs, but Cochran's overbearing personality also created problems. Steadfast in her views, she could be

Jacqueline Cochran poses in uniform at her desk as Director of the Women's Air Service Pilots in 1944 (photograph courtesy of Smithsonian Air and Space Museum, SI-2005-30504).

7. The WASP Experience

difficult to work with. In 1943, the Ferry Division lodged official complaints against Director Cochran. When two WASPs were dismissed on disciplinary grounds, and scheduled for an appeal with the Ferry Division, Cochran apparently pulled strings and removed the case from the division's jurisdiction.[8]

Occasionally Cochran's inspection visits resulted in controversy. Some base commanders criticized Cochran for making inspections without going through proper channels. According to the 2nd ferrying group, "On several occasions Miss Jacqueline Cochran has arrived at this base and gone directly to the WASP alert room, transacted her business, and departed without informing this Headquarters in any manner whatsoever."[9] Cochran challenged the charge and claimed that only once had she violated protocol. The issue was eventually resolved after a conference between Cochran and Training Command officers, Major General R.W. Harper and Lt. Colonel Joe Mountain. The latter requested that "Cochran's inspections be handled in the same manner as those of other members of the Air Staff."[10] Apparently, Cochran backed down and agreed to follow protocol.

At the very least, Cochran's actions disrupted the training command and strained cordial relations.[11] According to Cochran, her actions had been prompted by unfair treatment. If she announced her visits, the WASPs were mysteriously absent from the bases. Cochran believed this tactic was employed to keep her from interviewing the women. She also contended that surprise visits allowed her to see the base without fear of operations being "dressed up."[12] The commanding officers seemed less concerned about Cochran's surprise inspections than her refusal to announce herself upon arrival. Apparently, Cochran had been kept waiting once and was determined not to let it happen again. Since Cochran was often difficult and undiplomatic, it seems likely that both she and the training officers were to blame. But the situation was also partly created by the program's lack of military status. As a civilian, Cochran could not be ordered to comply with any military regulation. Her cooperation was strictly voluntary.

* * *

While women never made up more than 2 percent of the military during World War II, their presence released men for combat.[13] As air service pilots, women tested military aircraft, towed targets, conducted mock gas and strafing attacks, and, as ferry pilots, logged 70 million miles.[14]

One of the primary duties of WASPs was ferrying military aircraft from the factories to air bases. Women pilots flew 80 percent of all ferry missions in the war.[15] By assuming the role of ferrying pilots, women released hundreds of men for combat. The WASPS were stationed at ferry bases and sent out on daylight missions throughout the country.

The women's lives, as full-fledged WASPs, continued much as they had during training. Ferry pilots started their days at 6 A.M. and worked all day. The flying missions, however, were sporadic. According to *Liberty* magazine, "Their B-4 bags are always packed, because a woman ferry pilot seldom knows in what direction she will be pointing the nose of a plane an hour hence."[16] In between flights, women pilots continued to improve their skills. They studied navigation, radio communications and new flying techniques.[17] When an airplane needed delivery, the Ferry Command would send a woman pilot to the factory. Once there, she took the airplane up for a test flight and then delivered it to its destination.[18] According to former WASP Anne Dailey, "We were officially known as engineering test pilots. We had to check out any planes that had any major overhaul work before; they would let the men fly them, the women had to test it first."[19] The hours were long and the work was dangerous. Besides flying new planes to bases, WASPs flew damaged and experimental aircraft, delivered old airplanes to junkyards and worked with live ammunition.

Once the airplane was delivered, the women returned to the bases on trains, buses, or commercial airlines. A WASP held a "number two" priority rating on the airlines, which guaranteed her transportation back to the bases. If a WASP needed a seat, she could "bump" senators, congressmen, and even cabinet members, anyone except the president. The rule reflected the great need for ferry pilots. Once her mission was completed, the WASP hurried back to the base to deliver another plane.[20]

WASPs ferried airplanes from one end of the country to the other. The number of hours accumulated by WASP ferry pilots depended on the availability of airplanes.[21] According to WASP Barbara Poole, "I personally ferried approximately 400 hours in the first nine months. All the

7. The WASP Experience

girls worked just as hard and were proud of the fact that Ferry Division Headquarters at last decreed that the girls had earned the opportunity to ferry any airplane they were capable of handling."[22] In one specific ferrying mission, WASP Teresa James was ordered to ferry a P-47 Thunderbolt from Farmingdale, Long Island, to Evansville, Indiana. Upon reaching her destination, she was then ordered to take another P-47 to Long Beach, California. Once in California, she ferried a P-51 to Fort Meyers, Florida. In Florida, James was ordered to take an AT-6 to Tulsa, Oklahoma. From there, she ferried a P-39 Airacobra to Great Falls, Montana. Finally, in Montana, James was ordered to ferry a P-47 back to her base in New Castle, Delaware. In this one mission, James ferried six different airplanes in four weeks and logged 11,000 miles.[23]

The WASPs approached ferrying work with courage and determination. They saw themselves contributing to the war effort and took great pride in their flying. According to former WASP Margaret Chamberlain Tamplin, "When you delivered a plane and they signed for it and you knew that you had flown all across the United States, or I mean at least the worst part of the United States, it was a great feeling."[24]

At the start, women pilots were organized solely to deliver airplanes to air bases. The 25 WAFs organized by Nancy Love were ferry pilots and had no other duties. When the program was renamed the WASPs, a majority continued to perform this duty. Jacqueline Cochran, however, had other, more grandiose, ideas.

Unlike Nancy Love, Cochran wanted to see women's roles expanded beyond ferrying.[25] In July 1943, Cochran suggested that women pilots be trained to tow targets in the training of artillery and anti-aircraft gunnery.[26] Some 25 women were chosen for the special training, and their numbers gradually increased. After training, women pilots towed targets with three squadrons in the 1st Air Force and at least one squadron each in the Second, Third and the Fourth. They were considered support personnel for anti-aircraft and aerial gunnery training.[27] The women overcame the usual skepticism and established an accomplished record for this work.

Towing targets was a dangerous job, in which pilots pulled a 35-foot "flag" target connected to the airplane by a 300-yard cable.[28] Both the target and cable added extra weight to the airplane. WASP Mary

Ellen Keil recalled "how [the cable] would snap the plane up into a stall position and you had to be so careful when you released the target."[29] The work was also dangerous because it involved the use of live ammunition. According to former WASP G.C. Brown Kindig, "Towing targets was quite an interesting experience."[30] The artillery men that were shooting at the target frequently missed the target and we came down with holes in the plane, which was exciting to say the least." WASPs Madeline Sullivan and Sadie Hawkins had their engine shot out, forcing them to make an emergency landing.[31]

Occasionally, the green artillery men shot the airplane on purpose. In one towing mission, WASP Eileen Roach pulled a target from an A-24 when she heard cracking sounds against the fuselage. "Hey, I think you are hitting me!" she relayed on the radio. "Sure am sorry," came the reply, "Some gunner down here thought he was supposed to shoot at the plane."[32]

Despite the dangers, some WASPs wanted the position because it offered a variety of different airplanes and missions.[33] Women assigned to tow target squadrons also went on daytime tracking missions, nighttime searchlight-tracking missions, radio control work, and strafe and gas-attack missions. In the latter missions, women pilots flew over heavily camouflaged camps and "attacked" the American soldiers. As one WASP explained, "Peeling off, with the sun at our back, we'd dive down on emplacements, trucks, chow lines or anything visible. The men would track us with guns, simulating actual combat experience."[34]

In addition to training artillery men, WASPs were also used to help train bombardiers. They served as bomber pilots in airplanes with live bombs. Women piloted the airplanes as men practiced navigating and lining up targets. Former WASP Byrd Howell Granger recalled how one WASP landed an airplane with a bomb hanging from the bomb bay doors. Unable to release the bomb mechanically, the bombardier tried to dislodge it with his foot. When one tactic failed, the WASP took the plane in without exploding the shell.[35]

Jacqueline Cochran wanted women pilots to assume various flying duties. She believed if the women proved themselves as target towers, they could perform other jobs. According to Cochran, "We decided to try them [the WASPs] out on one of the hardest jobs the Army offered. We knew that if they were successful they could perform any of a dozen

7. The WASP Experience

other flying services for the Army, including instruction."[36] Shortly after the target towing experiment, Cochran suggested other roles for the WASPs.

One such role was the Weather Wing, which allowed weather pilot officers to be reassigned for combat.[37] In late 1943, ten women pilots completed training and began flying weather personnel to bases in Kansas City, Seattle, Santa Monica and Atlanta, and flew from airbases in San Antonio, and Long Island. Some WASPs, such as Babette Demoe, flew personnel to small islands in the Pacific, where radio communication

At the last WASP class graduation, director Jacqueline Cochran talks with Lieutenant General Barton Yount at Avenger Field, Sweetwater, Texas, in December 1944 (photograph courtesy of Smithsonian Air and Space Museum, SI-2005-30176).

could not be used.[38] The WASPs' overall performance inspired the Weather Wing to request more women pilots, and by the war's end WASPs were flying weather personnel five or six days a week.[39]

The success of the WASP program is exemplified by the women's low accident rate and low number of fatalities. As Cochran noted in 1944, "The fatalities to WASPs have been comparatively few and non-fatal accidents have been almost nil."[40]

While flying thousands of missions, women service pilots were involved in 402 airplane accidents.[41] In early 1944, *Flying* magazine reported that the WASPs did not have any non-fatal accidents, except for two sprained ankles.[42] Before the program ended, more serious injuries did occur. One involved a target towing accident caused by pilot Joan Whelan. Whelan and co-pilot Fran Smith hit a B-24 while practicing diving through their formation. Whelan's airplane took off part of the tail section of the other plane and put a large hole in the right wing of her plane. Then, in a hurry to take responsibility for the accident, Whelan quickly exited the airplane and caught her arm in the moving propeller. Whelan referred to her injuries as "nicks" but they were serious enough to require hours of surgery.[43] Despite this accident, WASPs who towed targets did not cause any fatalities, nor did any of their airplanes crash.

During the program's duration, 38 women were killed, with most of the fatalities occurring in the Ferry Division.[44] The cause of the fatalities varied from bad weather to pilot error to mechanical failure. Three women died while serving as co-pilots with male pilots. Another WASP was killed while riding as a passenger in the line of duty.

The first fatality occurred in 1943. The victim, Cornelia Fort, was a former flight instructor from Hawaii. On the morning of December 7, 1941, she was in the air near Pearl Harbor, but survived the Japanese attack. Fort, one of the original WAFs, joined the war effort in 1942. In the summer of 1943, she died in a bomber crash when a male pilot accidentally clipped her airplane's wing.[45] In 1944, WASP Betty Stine died in a training accident. On a flight to Blythe, California, her airplane, an AT-6, became disabled. Stine jumped from the plane and crashed into a rocky cliff. She died en route to the hospital.[46] In another mechanical failure, Dottie Nichols was killed when her P-39 engine quit on take-off. Fire engulfed the plane and Nichols was unable to escape.[47]

7. The WASP Experience

The death of a woman service pilot was tragic for several reasons. Not only did a young woman lose her life in the service of her country, but she received little recognition. The WASP program attracted little publicity, and the death of a WASP even less. For example, the report of Cornelia Fort's death was a postscript in an article she had written. *Flying* magazine did note that the "fatal accidents have occurred to pilots who seemed to have been best qualified to avoid them."[48] Since the WASPs were a civilian organization, its members received no recognition from the military. The body of a deceased WASP was simply returned to her family with no official ceremony. According to one grieving mother, it was as if the government said, "Thanks, here's your daughter."[49]

Some families were left with financial burdens. When a WASP died in the line of duty, her family received no death benefits or funeral expenses. By contrast, the government gave servicemen a $10,000 insurance policy, a casket, and transportation home. After WASP Evelyn Sharp was killed, her fellow service pilots took up a collection to send the body home.[50] Jacqueline Cochran personally paid for at least one funeral.[51]

Women service pilots showed great courage when one of their own was killed. In many ways they reacted as did soldiers in combat. While WASPs were shocked and saddened by these deaths, they refused to let emotions stand in the way of their duty. As individuals, the women had various reactions. WASP Mary Strok referred to the death of her best friend Betty Stine as "a shock." She also added, "But what can you do? It happens."[52] Meanwhile, WASP Adela Scharr criticized herself and Jacqueline Cochran for weeping at Dottie Nichols' funeral. She believes WASPs "should have been busy delivering [airplanes]," rather than mourning.[53] A more sensitive reaction came from WASP Elizabeth MacKethan Magid. Devastated by the death of her 19-year-old friend Marie Michelle Robinson, she wrote a tribute to all WASPs who had died in service. "She is not dead," wrote Magid, "but only flying higher; higher than she's flown before."[54]

* * *

Cochran's determination to expand the functions of WASPs enabled women flyers to perform many services for the war effort. At the same time, however, their progress was often hindered by gender discrimination. The WASPs confronted persistent hostility during their two years of service. As women, doing jobs traditionally defined as men's, they met skepticism, hostility and grudging admiration.

In 1943, the women were limited in the types of airplanes they could fly. A special board, set up to test and evaluate women pilots, concluded that WASPs "were not capable of flying the B-34 or the P-47, but could serve as co-pilots in the B-34 or the R-37 types."[55] The board did not believe that women had enough strength to fly the bomber.[56] Later the War Department issued a statement explaining the work of WASPs in two target squadrons, mentioning that the missions "are made in B-34s."[57] Despite the initial report, WASPs apparently proved themselves capable of flying the forbidden bomber.

The WASP ferrying program benefitted both the America war effort and the American Allies. When the Soviet Union needed fighter planes, WASPs ferried P-39s and P-63s to awaiting Soviet women pilots in Fairbanks, Alaska.[58] This idea was suggested by Nancy Love. The WASPs, however, were not allowed to fly the airplanes into Alaska. The Ferry Command feared women pilots might crash on a glacier, with no chance of rescue.[59] WASPs ferried the planes to Great Falls, Montana, where they turned them over to male pilots. Despite Love's complaints, the Ferry Command would not allow the women to fly the complete route.

Women pilots were limited by other gender barriers as well. Their roles were secondary and subservient to men and they were constantly told this is many ways. For example, WASPs were not allowed to instruct male pilots. In March 1944, Cochran suggested that women pilots "be assigned to basic flying schools to test the feasibility of using women to instruct men and to lighten the burden of a class of very much overworked pilots."[60] The women were rejected as flight instructors because officials felt men would not accept being taught by women.[61] The women did service as instrument instructors, teaching men and flight instructors among themselves. At the WASP training filed in Sweetwater, four women served as PT-19 instructors and one as a check pilot. Referred to as "sky-teachers" by the press, these women were an important part of the WASP program.[62]

While most WASPS encountered some form of hostility, ferry pilots, because of the nature of their duties, confronted the most discrimination. Unlike target towers, who only had to deal with hostilities of one squadron, ferry pilots hopped from one base to another, encountering many different circumstances. Ferrying aircraft was also a popular job with men, and they resented being replaced by women. According to one WASP, a skillful pilot with over 1,500 hours of flying time, the worst kind of discrimination was the silent treatment. As she recalled: "One day three of us landed some PT's at one airfield we prefer not to mention.... Nobody spoke to us from the time we landed until the time we left. We walked into the Operations office, checked in, and still we were met only with an air of hostile resentment."[63]

Ferry pilots also suffered from an identity crisis. The public, vexed by the ladies wearing silver wings, did not know who they were or what they did. In the early years, when the women were called WAFs, they were mistaken from everything from elevator girls to Girl Scout junior commandos, and even members of the Mexican Army.[64] An elderly Texas lady congratulated WASP Winifred Wood on her great work with the "jumping nurses."[65] This confusion caused problems for women pilots both on and off the bases. A group of parachute-toting WASPs were almost barred from a train because no one understood that they were working with the military. Meanwhile, another WASP was denied entry into a restaurant because she was wearing slacks. Unable to convince the person in charge that she was in uniform, she quietly left.[66]

On the whole, women were more easily accepted as target towers than ferry pilots. As Cochran had foreseen, since the male pilots disliked towing targets, they welcomed the presence of the WASPs.[67] The unpopularity of the duty served to benefit the women pilots and eased their acceptance. According to the chief of the Tow Target Section in the Fourth Air Force Headquarters, towing targets was an unpopular duty among male pilots. They considered the job a "dull chore" while most WASPs considered it a "high adventure."[68]

While most WASPs were welcomed as target towers, they were still subject to sexist criticism. In the official historical report, Lt. Daniel J. Stewart, a radio-control flight leader in a tow-target squadron at March Field, complained that the women were undertrained and "had no inclination to plan ahead."[69] He also contended that the types of flying and equipment used in Radio Control were "beyond the comprehension of the WASPs observed."[70]

Stewart's complaint seemed to revolve around one woman as opposed to the whole group. His (and other complaints) conflicted sharply with the official findings that the WASPs were indeed "capable of tow-target work."[71] According to official reports, the women "in many respects ... were better adapted to the activity then most pilots returned from combat."[72] This belief, however, echoed the discriminatory attitudes in factories. As William Chafe has pointed out, women were deemed better suited for dull, repetitive work than were men. According to the National Metal Trades Association, "Women are more patient, industrious, painstaking, and efficient about doing the same thing over and over again."[73] Towing targets was a military version of such work.

The initial reaction to a woman pilot was surprise, then skepticism, and finally, acceptance. When WASP Betty Gillies flew a P-47 into an Army airfield, one officer exclaimed, "Good Lord, do you see what I see? The pilot! It's a girl — and that's a P-47."[74] WASP trainee Kay Clewis had a similar experience when she landed at Coleman Army Base. As she prepared to climb out of the cockpit, two eager cadets greeted her with, "I say, sir, could we take a look at your plane." Their next response was "Holy gee, it's a girl!"[75] Another WASP, who landed at a Florida airport, was questioned by the tower, "Look here, you, why didn't you tell me you were a girl?"[76]

Once the women proved themselves in the air, they were more readily — if reluctantly — accepted. For example, when WASPs were assigned to the Second Air Force to ferry airplanes, tow targets and conduct radar and weather missions, their presence was received with misgiving.[77] According to one officer, "Many of the enlisted men assigned to the tow-target squadron did not relish the idea of 'flying with a woman pilot' at the time the WASPs arrived at Gowen Field, but after the girls had demonstrated the safe and sane piloting procedures that proved characteristic of the group, the original skepticism of the enlisted men turned to admiration."[78]

Jacqueline Cochran herself noticed and commented on this change of attitude. In August 1944, she inspected between 18 and 27 bases in 24 days. She met with 200 WASPs and investigated their activities and well-being. According to Cochran, when she visited ferrying bases in June, the commanders treated WASPs with indifference or, at best, minimal toleration.[79] When she returned in August, she discovered a new attitude among some of the commanders. Cochran recounted that one commander requested ten more WASPs in August, a decided change from his earlier request to reduce the number. Meanwhile, the Second Air Force requested over 200 women pilots.[80] Cochran, surprised at how quickly attitudes had changed, concluded that the WASPs had "had long enough to sell themselves as a contributing unit to the Air Force."[81]

WASPs were keenly aware that their presence disturbed some of their male co-workers. The women referred to their skeptics as "NBs," or non-believers.[82] The discrimination, however, seemed to be a secondary concern. Women pilots loved flying and helping the war effort. They seemed to view discrimination as something that could be expected and had to be accepted. According to one WASP, "Until these NB's will accept us as pilots and not resent us simply because we are women our jobs are made just that much harder." She added, "Until this war is won those of us who can fly are entitled to feel that we are making our own kind of contribution toward that victory."[83]

The press practiced a more subtle form of discrimination and gender inequality. While praising WASPs for doing their part, the press reinforced society's view of women workers. Reporter John Stuart of *Flying* magazine referred to WASPs as intelligent, efficient girls who "are setting an enviable record as civilian employees of the Army Air Forces."[84] He also pointed out the "frilly" decorations in the WASPs' quarters which included "lacy pillows, silken coverlets and modern novels."[85] An earlier article in *Pegasus* magazine, told how WASPs are going "about their work with a totally ungirlish drive." However, the author notes that the women are very feminine, have slender waists and hips and when marching they sing "very very lady-like songs."[86] Most articles pointed out that the WASPs were strictly temporary. After all, the WASPs, like women in industry, were learning new skills to advance the war effort, not redefining themselves. Women might be riveting, welding and flying military aircraft, but their new roles were temporary. According to Frank Taylor of *Liberty* magazine, "The women warbirds know they are flying as an organized group on a wholesale basis for the duration only."[87]

The press portrayed WASPs, much like women working in industry, as glamour girls. The glamour theme, as Ruth Milkman has noted, reassured the public that "war work need not involve a loss of femininity."[88] Reporters worked diligently to assure the public that the women pilots were still women. Even women journalists like Charlotte Knight, who frequently spotlighted the WASPs for *Air Force* magazine, wrote fluff pieces. According to one Knight article, "The frills are gone but flying hasn't interfered with feminine grooming."[89] Meanwhile, other reporters referred to the WASPs as "feminine fliers," angels, or direct threats to Adolf Hitler. According to one article, "These girls ... their hearts set on piloting with an unfeminine purpose that might well be a threat to Hitler," the implication being that any country that resorts to using women must be intent on winning.[90] In another article, they were called "a subtle threat: to the Axis."[91] "There's a subtle charm — and, for the Axis, a subtle threat — in the appearance of so much femininity in an activity about which there is such grimness and determination."[92] These attitudes trivialized the program, while also reinforcing notions of sexual difference.

The WASPs' duties were also described in feminine terms. Charlotte Knight's article, "She Wears a Pair of Silver Wings," attempted to explain that the job of towing targets was "no tea party."[93] Another article, discussing the WASPs' mission preparations, admitted they talk a lot, but "its not tea-time conversation."[94] Meanwhile, *Time* magazine reported that WASPs flying trainers in bad weather "sweat it out with womanly patience."[95]

Why were women pilots portrayed in this sexist way? And why did the WASPs not protest the treatment? The 1940s was a very different decade from today, women were viewed as having a domestic role. The fact that women were taking over jobs formerly performed by men and doing those jobs well surprised many Americans. In the WASPs' case, women did jobs they had never done before. According to former WASP Adela Riek Scharr, "Some of them no larger or stronger than a 13-year-old boy proved themselves capable of matching the men's aerial performance."[96] They also took over jobs that men enjoyed, which caused additional resentment. As Scharr explained, having women in civilian flying roles "forced [men] to go to war instead of remaining sadly at home."[97] To help counter this resentment, as well as explain women's success, the difficulties of the job were played down and femininity of women played up.

Women pilots endured the sexism and resentment without any formal protests. Sometimes they sympathized with the men they replaced. According to former WASP Jean Hascall Cole, "The men ... resented being moved out of a safe job and forced overseas to join the fighting. Who can blame them? It was a long and deadly war."[98]

WASPs dealt with the bad times by forging strong friendships with one another. "One big plus was the comradeship of the women," explained Cole. "To this day classmates remain friends."[99] This factor, coupled with the love they felt for the job and the patriotism of helping the war effort, helped them endure the inevitable criticism. "They were a marvelous group," commented former WASP Mary Strok, "all loving one single thing—flying."[100]

Women also endured the unequal treatment out of a sense of patriotism and for the opportunity to fly. According to former WASP Betty Jane Williams, "We did something for our country because in those days women were doing everything to advance the cause of the war.... If you were not working in the red cross making bandages or Rosie the Riveter you were doing something."[101] Unlike factory work, flying airplanes was an exciting, even dangerous, challenge. The women in the program were already pilots and eagerly accepted the chance to fly. According to Williams, "This was wonderful, to do this for your country and do something that you loved."[102]

Throughout the program's duration, Jacqueline Cochran both challenged and reinforced gender roles. As WASP director, she took on the conflicting tasks of expanding the role of women in military aviation, while reassuring the public of their femininity. It was a task for which Cochran was well suited. Since 1935, she had been blending the roles of aviator and cosmetics queen. Cochran, a pioneer in the masculine field of aviation, maintained a decidedly feminine image. Now she used her glamorous image to advance the WASP program.

As the war began to wind down and men returned home, women military pilots were removed from their positions. During the war, Director Cochran had assured the public that the women would not mind leaving the WASP program. "I want all my girls, those at Sweetwater now and those who will follow them later, to raise families and make stable homes for our men when this war is over. And I know that's what they want too."[103] The transition was easier for some women than others. According to former WASP Fran Laraway, after the program was deactivated, "I was just in quite a lull, where I watched the sky and missed it very much."[104]

* * *

As director, Cochran's responsibilities were strictly administrative. She neither trained nor directly supervised women pilots. Cochran, however, did successfully expand the program and the role of women in the military.

Part of the key to Cochran's success was her willingness to conform to traditional standards. For example, she helped reinforce gender discrimination by referring to the WASPs' duties as "aerial dishwashing." She willingly used gender stereotypes to help the WASPs gain acceptance. Cochran knew that men would view women pilots with skepticism and hostility, but preferred to work within the system, rather than change it. "I steered them into work that men didn't want to do and wanted no part of it," Cochran commented in later years.[105] Since men enjoyed the ferry work, she "made girls do the other chores, the dishwashing."[106]

Cochran's choice of words implies that some of the WASPs' duties were akin to monotonous household chores, duties suited to women. Although she expanded the WASPs' duties, she believed that women should be content to do the jobs rejected by men. Cochran's intent was not to belittle women, but get their foot in the door, by any means. Yet, in this case, she reinforced the idea that women could only understand jobs when related to housework. Although a woman, Cochran used the same kind of terminology as male employers in industrial positions. Attempting to explain the success of women factories workers, some employers claimed that operating an overhead crane is just like using a gigantic clothes wringer or "the winding of wire spools in electrical factories [is] very much like crocheting."[107]

Cochran pursued two major objectives: to create greater roles for women pilots and to maintain traditional roles. She pursued these, in part, by carefully selecting who would be in the WASP program. Fearing bad publicity, Cochran was determined to create a spotless organization. She purposely selected women not likely to embarrass the program. "I didn't let anybody get into this program that was not from a good, moral, serious background."[108] To ensure a good candidate, Cochran checked an applicant's family background and had intelligence people conduct other background checks.[109]

Cochran also chose to avoid controversy by upholding racial segregation. When a qualified black woman applied to the program, Cochran quietly persuaded her to withdraw.[110] He decision to block black applicants was inspired by her own conservatism, as well as the times. Until 1941, black men could not join the Army Air Corps. After 1941, when the ban was lifted, black men were segregated from white men, as they were in all branches of the military. Cochran could not establish a black WASPs Corps with a single pilot. Hardly a social crusader herself, Cochran would not risk attacks on the WASPs by breaking the color barrier.

Cochran used her personal image and fame to generate good public relations. Even in controversy Jacqueline Cochran was good press. She was frequently referred to as "smart," "pretty," "glamorous," and an "outstanding" or "famous pilot" and even "mother" of the Sweetwater training facility.[111] Cochran could always be counted on for the right quote. She continually stressed the temporary status of the WASPs and their place in postwar society. "Women will never make peacetime transport fliers," she explained in 1943. "I know they won't play an important role as fliers. But there's a job they can do today and I want them to get the chance they deserve."[112]

Jacqueline Cochran created a role for women flyers in the war, but did her actions help or impede the cause of women? Allowing women to fly military aircraft as a break with tradition had the potential for permanently altering gender roles in both the military and aviation. But when the war ended, so did the military flying career of women. Commercial aviation also remained closed to women pilots. After the WASP program was disbanded several airlines sent WASPs stewardess applications. The women reacted to this discrimination with sadness, amusement and anger.[113]

To what extent can Cochran be blamed for impeding the advancement of women in aviation? After all, she followed traditional views and did little to challenge gender roles. In fact, Cochran often reinforced gender roles by stressing the femininity of WASPs. To call Cochran an obstacle, implies that she was a barrier to some movement, when, in truth, none existed. After the WASPs disbanded, women pilots did not protest their release. They did not picket Air Force bases and airports, or march on Washington. Instead, they returned to their homes, as expected.

When Cochran organized women pilots she was meeting a wartime need, not attempting to redefine social norms. Her views were shared by many women in the program. According to one WASP, "When [the war is] over, we can go back to whatever we were doing before.... For myself— and for a number of us — that means going back home and caring for my two children."[114] "Back then things were really quite different from what they were now," explained a former WASP in 1990. "It wasn't an easy adjustment [trading flying for housework], as many suggested, but it was accepted, and I suppose for many of us that was how it had to be!"[115]

Cochran had the ability and opportunity to launch a movement for women, but her attention was elsewhere. She, therefore, can be blamed for not leading a crusade to advance women in aviation. An outspoken woman with powerful connections, Cochran had much to offer such a campaign. In fact, she possessed many advantages to boost a crusade for women — a famous name and a very rich husband. She was an internationally known celebrity with access to millions. Women's rights, however, held no fascination for her. She was more interested in what she could do for herself and for aviation. After an unsuccessful attempt to militarize the WASP program, Cochran had it disbanded. Leaving the cause of women behind, she traveled to Asia and Europe to witness the end of World War II.

8

The End of the WASPs

The desire to fly fast aircraft couldn't be quenched. It would be long before I must go home, but I could not tear myself away as readily as others had done.[1] — WASP Adela Riek Scharr

In a November 1944 article, WASP Director Jacqueline Cochran praised the accomplishments of America's first women service pilots. "Two and a half years ago," she commented, "some of us thought American women had strength and stamina enough to be of tremendous potential use in the air, but nobody really knew. We had all heard tales of the Russian women flying heavy planes and doing parachute jumps for fun, but nobody knew whether ours could do it too. I thought so. It has been an extremely interesting experience to watch them develop into topnotch pilots."[2]

Since 1942, women pilots had been supporting the Army Air Force. As previous chapters have shown, these women overcame deep skepticism and were soon performing jobs once held only by men — flying bombers and fighters, testing and delivering planes, towing targets for gunnery recruits and strafing troops in training camps.

The WASP Program gave the women pilots a chance to aid the war effort by employing their greatest skill — flying. Possibly, the program could have advanced women pilots in the postwar world. After all, women had proven themselves as capable, skilled aviators. Surely there would be a place for them after the war ended. Even Jacqueline Cochran believed WASPs would continue to fly. In an interview with the *New York Times* she said: "You won't keep these women out of the air. People never stay grounded once they learn to fly."[3]

But Cochran was wrong. A few weeks after the interview, and five months before VE Day, the WASPs were disbanded, and a few of the women service pilots would ever fly again.

* * *

There were several issues involved in the disbandment of the WASPs. One was the quest for military status. The WASP program had been organized as a civilian agency, women pilots had neither military rank nor benefits and were not subject to military discipline. Initially, Cochran had objected to allowing women pilots to have military status, believing that the women needed to prove themselves first. She had also selfishly opposed incorporating women pilots into the Army's Women's Army Corps (WACs) because she knew if women pilots were placed under the WACs, a WASP program would not be formed and she would never become its director. In 1943, when Cochran was now firmly in charge of the WASP program, she had a change of heart and supported militarization.

The WASPs were the only auxiliary military group who did not enjoy military benefits. Yet, despite its civilian status, the WASP program was run as a military operation. The women wore uniforms, conformed to military rules, and performed military duties. According to Director Cochran in her final report, "They were living at AAF bases dealing with Air Force Equipment, eating in officers mess rooms, and associating with flying personnel, and yet were governed by an entirely different set of laws and regulations."[4] But, as civilians, the women pilots were ineligible for military benefits such as hospitalization, medical care, funeral expenses, retirement or the G.I. Bill. Militarization was a just cause. The women may have been home-front "warriors," but they were still risking their lives in the service of their country.

In 1944, when the issue arose again, several powerful people were on the WASPs' side, including Secretary of War Henry Stimson and Air Force Chief General Hap Arnold. In January 1944, Arnold wrote Chief of Staff George C. Marshall, urging support for WASP military status.

"[The WASPs] are performing a valuable function," Arnold wrote. "They are taking flying risks without the benefits and protection available to male personnel. Matters of hospitalization, insurance and protection of dependents require that they be placed in equivalent status. Administration will be greatly simplified."[5]

8. The End of the WASPs

Several members of Congress also supported military status including Democratic Congressman John Costello of California who, in February, introduced a bill providing for appointment of women pilots as officers in the Army Air Forces.[6] The women would be given rank from flight officer to colonel, although Cochran would be the only colonel. They would also receive the same pay, privileges and benefits as male Army Air Force officers.[7] Arnold publicly supported the proposal and, in March, testified before the House Military Affairs Committee. He continued to praise the women pilots and emphasized the importance of their work. Stressing the "serious manpower shortage," Arnold pointed out that "it is a problem that affects the fighting forces of the United States, and we must provide fighting men wherever we can, replacing them by women wherever we can; whether that be in the factories, ferrying aircraft across the country, towing targets for ground troops to shoot at, or any place where we can release men and make available the younger men to actually do the fighting."[8] "The WASP Program," continued Arnold, "has been found to be military sound and necessary. Women pilots are available and qualified for certain flying duties which they can discharge as competently as men."[9]

Despite Arnold's support, members of Congress questioned the necessity of the WASP program. After all, women pilots were intended as temporary replacements because men were not available. In early 1944, this situation began to change. The Army and Navy discontinued the war training service program under the Civil Aeronautics Administration, which employed some 900 pilots and another 4,167 student instructors. According to the *New York Times*, "Many of these men will lose their deferred draft status and may be put into the "walking army."[10] Rather than join the infantry, these men wanted the flying jobs currently held by WASPs. Another 10,000 male pilots from the Civil Air Patrol could also be used as service pilots. Secretary Stimson defended the WASPs and assured critics that "neither the existence nor the militarization of the WASP will keep out of the Army Air Forces a single instructor or partially trained civilian pilot who desires to become a service pilot or cadet and can meet the applicable standards of the Army Air Forces."[11]

Stimson, however, did not convince Congress of this. To determine the usefulness of the WASPs, Congress authorized an investigation by the House Committee on Civil Service, headed by Robert Ramspeck of California. Ramspeck's Committee concluded that the WASP training program was a waste of money.[12] According to Ramspeck's report, the "Army Air Forces had embarked upon a costly and unnecessary program of recruiting inexperienced young women for training as noncombatant service pilots and the Army Air Forces as dismissing, or failing to properly utilize, large numbers of male civilian pilot instructors, who had been trained at a cost of millions of dollars."[13] Ramspeck concluded that the need to "recruit teenaged schoolgirls, stenographers, clerk, beauticians, housewives and factory workers to pilot the military planes of this Government is as startling as it is invalid; the militarization of Cochran's WASPs is not necessary or desirable: the present program should be immediately and sharply curtailed."[14]

Director Cochran wrote her own report contradicting the Committee's findings. Despite Ramspeck's stinging comments, her response was uncharacteristically diplomatic. In a letter to Congressman Ramspeck, Cochran said she was writing to "remove ... [Ramspeck's] critical or disparaging thoughts" about the WASPs. "[The WASPs] are working and dying in the air alongside our male pilots, are very close to my heart and ... deserve better treatment than accorded them in the report." Cochran also expressed sympathy for the unemployed civilian male flying instructors, whose plight made the WASPs seem unnecessary. "The WASPs," wrote Cochran, "were struck at not to hurt them but to help these civilian instructors. When we stop to consider that there were several thousand male civilian pilots who believed their situation urgent and only about 60 WASPs were being added to the AAF working forces each month we see how the continuance or stoppage of the WASP training program could not affect the ability of the Air Force either to absorb or not to absorb quickly these civilians.... In my humble opinion, none of this had any bearing on whether the WASPs actually in service should be militarized."[15]

The WASPs greatest obstacle was their gender. As women, they were expected to aid the war effort only on a temporary basis. Despite the fact that the WASPs were talented, well-trained, capable flyers, their existence would not be tolerated if male pilots were available. On the House floor, the Costello Bill was severely criticized. Democratic Con-

8. The End of the WASPs

gressman James Morrison of Louisiana referred to the WASPs as "perhaps the most super-duper of all programs" and implied that the women pilots were only glamour girls in designer clothes. "Magazines have played them up and even the movies contributed a picture in their behalf," he criticized. "Their natty and stylish uniforms were tailored on Fifth Avenue in New York and cost over $500 for each WASP."[16] According to Morrison, if the WASP program continued, experienced male pilots "may soon be cleaning windshield[s] and servicing airplanes for glamorous women fliers who have only thirty-five hours of flying time."[17]

Republican Congressman Joseph O'Hara also opposed the WASP bill because men were available to fly. Referring to the WASPs as "charming young ladies," he conceded that "some of them are very able and very fine pilots." But the Army, he lamented, would have to tell qualified male pilots to "become grease monkey[s] or tail gunner[s] ... because his natural ability does not fit into this program, because we want to have the girl pilots." According to O'Hara, male pilots will be turned away because "we have to have somebody in there who is a very attractive lady pilot, who is needed in the program."[18] Democratic Congressman Edward Izac of California described the WASP bill as "the most unjustified piece of legislation that could possibly be brought before the House." He also believed that granting military status to the WASPs would not really change their situation. "No change," Izac claimed, "except the change that comes by putting on a manikin's uniform."[19] Military status for the WASPs, Izac contended, is "not going to help the course of the war one bit, and I am sorry to see Hap Arnold lose his balance over this proposition."[20] Part of Izac's resentment can be traced to his connection to the Navy. Regarded as a "Navy Spokesman," Izac was also a member of the Naval Affairs Committee, an Annapolis graduate, and Congressional Medal of Honor winner for his service in the Navy during World War I. Izac resented spending money on programs such as the WASPs when that money could have been appropriated for the Navy. He also wanted to ensure a place for Navy Air Cadets after the Navy reduced its pilot-training program.[21]

Some Congressmen supported the WASPs. Andrew May, Kentucky Democrat and Chairman of the Military Affairs Committee, favored the bill, as did Republican John Vorys of Ohio. The latter, a licensed pilot, had recently experienced a change of heart about the WASPs' capabilities. According to the *New York Times,* Vorys "described his reversal of opinion on the skill of the women pilots after he had been tested for flying ability by one of them and had made the rounds of several air fields where they are serving."[22] The bill's sponsor, John Costello, defended the WASPs and assured Congressman O'Hara that the bill had nothing to do with "creating some glamorous organization or social organization." In an elegant rebuttal he challenged O'Hara's definition of glamour. "If you like to be covered with grease," Costello explained, "if you like to sweat out piloting an airplane through stormy weather from one coast to another and call that a social activity, very well, then vote against the bill. But if you want to try to help carry on the war effort, if you want to release a few more men to do flying in combat zones, to engage the enemy themselves, then pass this bill and release 2,500 pilots who otherwise will have to stay in this country."[23]

Sadly, Costello's bill was voted down by a vote of 169 (for) to 189 (against). A similar Senate bill proposed by Senators Lister Hill of Alabama and Harold Burton of Indiana never came to a vote.

Resentment of Jacqueline Cochran was another reason for the bill's defeat. Congressman Izac vehemently opposed making Cochran a colonel, even though the head of WACs, Oveta Hobby, currently held this rank.[24] He concluded that "any woman would like to have 2,500 girls under her and be a colonel ... for 2,500 WASPs we want to make some woman a colonel. That is just one of the sidelights of this thing."[25] Even Cochran felt her proposed promotion was one reason for the bill's defeat. "What it comes down to really," Cochran told WASP Byrd Howell Granger, "is that as long as I am Director of Women Airforce Service Pilots, WASPs will never be militarized." According to Granger, a teary-eyed Cochran asked Granger to take over her position. Granger refused. "No matter what happens," she assured Cochran, ["whether WASPs sink or swim, without you there can be no WASP. It is your baby.] If we go down, so be it."[26] Clearly, some men resented Cochran because she was a woman. She was outspoken, she saw a position of power, and prompted women to take jobs that belonged to men. Between her wealthy husband and close relationship with the Army Air Force, Cochran's influence seemed too great. According to Drew Pearson of the *Washington Post,* the Government "has spent more than $21,000,000 training lady flyers, primarily at the behest of vivacious aviatrix Jacqueline Cochran, wife of financial magnate Floyd Odlum. Magnetic Miss Cochran seems to have quite a drag with the 'brass hats' and has even persuaded the Air Forces' smiling commander to make several secret trips to Capitol Hill to lobby for continuation of the pets, the WASPs."[27]

As historian D'Ann Campbell has pointed out, while some "350,000 women were members of the U.S. armed forces [during World War II] their numbers represented not so much a triumph of feminism as a reflection of the military's manpower shortage."[28] When the shortage of male pilots ended, so did the WASP program.

The defeat of House Bill 4219 was the beginning of the end of the WASP program. Five days after the Congressional vote, General Arnold announced the discontinuation of all WASP training. Women pilots then currently in training would be allowed to graduate, but no additional applicants would be accepted.[29]

In August 1944, Director Cochran made a final appeal for militarization. In a report to Hap Arnold, Cochran again suggested that the WASP program be reevaluated and its status changed.[30] She also suggested a compromise. Cochran proposed that WASPs only be given the flying jobs that men did not want. She suggested transferring 126 from ferrying duty to target towing and other training operations "that most men dislike." She also suggested that women pilots fly only light bombers and leave the heavier combat type airplanes for the men.[31] In her closing statement, Cochran recommended "that serious consideration should be given to inactivation of the WASP program if militarization is not soon authorized."[32] Citing a "lack of control" over civilian employees, Cochran believed the program would not be effective unless militarized and, thus, would be better off dissolved. Even if deactivated, Cochran believed that "an effort should be made to obtain military status, if only for one day," so the WASPs could obtain benefits.[33]

Cochran's report did not benefit the WASP Program. By suggesting deactivation, she implied that women pilots were no longer needed. Cochran also devalued WASP performance by pointing out that a civilian program could not be as effective as a militarized organization. The text of Cochran's report was made public on August 8th. According to the press release, Cochran proclaimed that "on a long-term basis, dependability, control, and full effectiveness can be obtained only through militarization by incorporation into the Air Corps."[34]

Cochran's suggestion that the WASP program be terminated puzzled both the WASPs and historians. Sally Keil, in *Those Wonderful Women in Their Flying Machines*, saw Cochran's suggestion as an ultimatum to militarize the WASPs or else.[35] Former WASP Byrd Howell Granger, however, contended that Cochran's report was not a call for overall disbandment. She was simply "bouncing off" ideas for the WASPs' future.[36] Molly Merryman, in *Clipped Wings*, agreed with Granger, but failed to explain what Cochran meant. According to Merryman, the press twisted Cochran's words, making her suggestion sound like an ultimatum.[37] Another former WASP, Adela Riek Scharr believed the root of the problem was the jealous rivalry between Cochran and WAC commander Col. Oveta Hobby. According to Scharr, Cochran had been offered a compromise of being made a captain in exchange for passage of the bill. Cochran rejected this because she wanted to be a colonel. Meanwhile, "Hobby demanded of the army that if Cochran were to become a colonel, Hobby should outrank her not only because of the seniority in service but for having many more thousands of troops under her command. That declaration," concluded Scharr, "was enough to keep the bold ladybird [Cochran] from getting into the top echelon roost."[38] Scharr places the disbandment of the WASPs on Cochran's selfish ambition. She held out for militarization on her terms and lost.

Cochran had erred in siding against military status when the WASP program was first formed. She did not, however, give Hap Arnold an ultimatum when she suggested that the WASP be militarized or disbanded. Despite Cochran's brassy personality, she would not have dictated terms to General Arnold, a man she revered.

Secondly, Cochran was very proud of the WASP program and did not want to see it end. By August, however, she viewed militarization as a lost cause. For the first time in her life, she had hit a brick wall. Her great wealth and powerful allies could not help the WASPs' cause. Even her stylish manner, an asset in the past, was being used against her. When critics lambasted the WASPs, they usually included some reference to "pretty" or "vivacious" Jacqueline Cochran and how she was using her

8. The End of the WASPs

femininity to influence Hap Arnold. By suggesting militarization or disbandment, she was making a last stand.[39]

Sadly, disbanding the program was a simple solution to the WASP problem, and a few months later the Army Air Force took the easy way out. In October, General Arnold ordered the end of the WASP program, and all the women were to return home by Christmas. The last WASP training class graduated on December 7, 1944. Both Jacqueline Cochran and Hap Arnold attended this final ceremony and addressed the women pilots. "My greatest achievement in aviation," Director Cochran said, "has been the small part I have played in helping make possible the results you have shown."[40] In a letter to all WASPs, Arnold explained his decision about disbandment.[41] Contradicting his remarks to Congress, he said that male pilots were now available to fill civilian flying positions. "Now the war situation has changed," he wrote, "and the time has come when your volunteered services are no longer needed. The situation is that, if you continue in service, you will be replacing instead of releasing our young men. I know that the WASP wouldn't want that."[42]

General Henry "Hap" Arnold had been a staunch supporter of the WASPs and his decision to end the program shocked the women pilots. He had worked hard to attain military status, but was unable to overcome Congress's opposition. Arnold had also suffered several heart attacks and was no longer well enough to continue the fight. He had given up on the WASP military status and took Cochran's advice to disband the program. As Christmas approached, Arnold felt he was doing women pilots a service by allowing them to return home for the holidays.

For Cochran, the WASPs sudden disbandment was an unfitting end to a program she had worked so hard to establish. "It was with deep regret that I took on a task officially ordered by General Arnold" she wrote when submitting a plan for deactivation.[43] The publicity generated during the fight for mobilization had taken Cochran by surprise. She recalled how the WASPs were referred to as "teenage schools girls" or "my pets ... putting male pilots on bread lines," and lamented that these characterizations had overshadowed their excellent work.[44] Cochran, however, accepted Arnold's decision without a fight.[45] In a December 1944 press release, she, too, explained that the WASP program was no longer needed. "The WASPs are being demobilized," Cochran announced, "because the pilot shortage which existed in 1942 and which caused the woman pilot program to be started has now been effectively corrected. If continued in service longer, the WASPs would be using flying time on domestic operations that could be spread to better advantage among the available male pilots who are awaiting their assignments aboard or who have returned from their completed combat missions and are available for domestic duties. Also, the reduction of flying training has released for domestic flying from the civil contract schools many qualified pilots who had been instructing cadets. Thus the WASPs have successfully fulfilled their main objective."[46]

* * *

The sudden end of the WASP program both surprised and saddened its members. "It really took the heart out of a lot of us," explained former WASP Anne Dailey Marshall, "because we had looked forward to doing more than what we were doing. My grandfather was killed by Germans, one uncle was killed by Germans and I couldn't understand why they were disbanding us while the war was still going on." Many WASPs felt a sense of betrayal. "We were left with a feeling that something we had done was wrong," recalled Byrd Howell Granger, "that wasn't the case at all, it was a political thing."[47]

The WASPs sudden disbandment (without military status) left women pilots feeling abandoned. They had lost the opportunity to fly for their country and help end the war. According to WASP Byrd Howell Granger, "To be dismissed, discharged, at Christmas time with no transportation ... no discharge pay, nothing, just to be turned lose ... we were all heartsick."[48] Fellow WASP Alyce Stevens Rohrer agreed. "When the orders came," she commented years late, "we were devastated."[49]

Women pilots were disappointed for several reasons. First, they wanted to aid the war effort as long as the war continued. And secondly, they enjoyed their jobs and wanted to keep flying. According to WASP Barbara Erickson London, "This was the best time of our life. We never had a period in our life that gave what those three years gave us. We were at the right place at the right time. We were able to fly all of these marvelous airplanes. We were helping with a war that we were very serious about and we were very concerned about our country. And, this was probably the greatest experience in our lives for all of us."[50] WASP Winfred Wood agreed. "They took away my silver wings," she wrote, "but they left me with something brighter, something that won't tarnish until I am old and feeble and can no longer remember fun."[51] Very few of the female pilots would ever experience such excitement again.

* * *

After disbandment was announced, WASP pilot Betty Tackaberry was presented with two choices: leave her P-51 or fly it to Newark, New Jersey, then hop a flight to Long Beach, California, where the WASPs would be disbanded. She chose the latter. Sitting tenth in line to take-off, she began to have second thoughts. She eyed the ten-foot drifts of snow on each side of the runway and dreaded the icy runway at her next stop, Cleveland. She then began thinking about the last flights others had made: including Amelia Earhart. "People who said they were on their last flights and it turned out to be." Tackaberry turned the plane around and climbed out of the cockpit. "I shed a few tears," she recalled, "and walked away. I knew I would never be in one again. It was very sad."[52]

Like Betty Tackaberry, no WASPs ever flew military airplanes again. Few continued to fly at all, a fact that dispelled the 1944 prediction made by Jacqueline Cochran. In reality, few positions existed for women pilots. "The women who applied to airlines, trying to fly, were not allowed in," explained WASP Leona Golbinec.[53] "I tried the airlines," said WASP Teresa James, "and they all told me the same thing — public opinion wouldn't permit a female pilot in the cockpit, even if she had a four-engine rating."[54] WASP Ruth Adams tried to get a job as a commercial pilot for Republic Steel, but was turned down because the company president did not "think the officers of the company would be comfortable with a woman pilot."[55] WASP Helen Richey moved to New York and tried unsuccessfully to be hired as a pilot, flight instructor or flight consultant. Unable to resume her greatest passion — flying — Richey committed suicide in January 1947.

Unlike Cochran, the women in her program were not wealthy or famous. They would not have the means or the encouragement to continue flying. Not even private flying was an option for many women.

Airplanes were expensive and flying was too time-consuming for full-time wives and mothers. According to WASP Anne Berry, private flying "was still twelve dollars an hour, and I never had enough money to fly. Then I married and had children, and it just never seemed to work out."[56] Some former WASPs were content just to work near an airfield. WASP Anne Craft joined the Red Cross and served American soldiers coffee and cake from an airstrip hanger in Puerto Rico.[57]

WASPs who did continue to fly considered themselves lucky. Former WASP Madeline Sullivan became a copilot for a feeder airline with only one plane. Other former WASPs took up barnstorming or went to foreign countries like South America and flew "wherever they could find a little war going on." According to WASP Fran Laraway, some of the women "had learned to live on the high edge of things [and] couldn't bring themselves down."[58]

During the post–World War II era, Jacqueline Cochran resumed completion flying. Here, she is sitting on the left side of the cockpit sill of a North American P-51B Mustang, circa 1946 (photograph courtesy of Smithsonian Air and Space Museum, SI-82-13215).

Jacqueline Cochran's postwar activities contrasted with her WASPs pilots. She quickly resumed her flying career and put the WASP program behind her. She even gained personal military status when she was made a lieutenant colonel in the Civil Air Patrol and held the same rank in the Air Force Reserve.

After World War II,

8. The End of the WASPs

Cochran hoped that women pilots would keep flying. Yet, despite her money and influence, she did not launch a campaign for women flyers or challenge barriers that barred women from military and commercial flying. Cochran, as always, followed the convention of the day. After World War II, women were expected to return home and be good wives and mothers. Cochran also believed that a woman's first responsibility was her home and family. Therefore, she did not encourage women pilots to make a living by flying.

Cochran's own actions did not always follow her beliefs. After the WASP program was disbanded, she remained on General Arnold's staff and traveled to the Pacific at his request.[59] In 1945, Cochran viewed bomb-damaged Japan and then traveled to Germany. Here she toured Adolf Hitler's underground bunker, taking home one of the fuehrer's doorknobs as a souvenir. Cochran went on to attend the Nuremburg trials, during which she observed the sentencing of Nazi war criminals.

When Cochran returned to the United States, she lent her name and reputation to causes that were important to her. She became a spokeswoman on the importance of air power and joined the political fight to separate the Air Force from the Army. In 1947, a separate Air Force was created. Later, she crusaded for General Dwight Eisenhower to run for the presidency. In 1956, she ran for a seat in the United States House of Representatives.

Despite the WASPs' performance, women pilots could not fly for the military until the early 1970s. Women pilots flew first for the United States Navy and, in 1977, the Air Force. Their roles were limited. Women flew only certain types of aircraft: weather, reconnaissance, tanker, personnel and cargo transport and flying hospitals. In the 1980s, women pilots started flying Airborne Warning and Control Squadron (AWACS) and electronic countermeasure aircraft.[60] In the Gulf War, women pilots were barred from combat, but flew helicopters and transport planes in war zones.[61] Meanwhile, women pilots in the Navy are trained to fly fighter planes.

While Cochran often praised women pilots, she would not have approved of their role in the Gulf War or their participation in Iraq or Afghanistan. Cochran did not believe that women should be in combat or combat-like situations. Nor did she support women being in military academies. "I think it's morally wrong," she admitted. "I see no justification for it whatsoever, and I testified that before Congress, I think it's a fiasco."[62] When Cochran organized the WASPs in 1943, she stressed their roles would be temporary and limited. The WASPs' purpose was to aid the war effort and release, not replace, men. After the war, Cochran wanted the women to return home, not continue in the military. "I [am] against women in the military in peacetime," she commented in a 1976 interview. "I really am."[63] Therefore, Cochran's legacy for female military pilots was unintentional, and even ironic.

As America's premier female aviator, Cochran proved that women could fly as well as men. Her WASP program further elevated women pilots. Women proved that they were capable of flying all types of aircraft, from trainers to fighters and heavy bombers. After the war, women pilots could have continued flying for the Army Air Force. At the time, however, views on gender impeded their progress. Women were not welcome in the military, much less military aircraft.

* * *

During World War II, Jacqueline Cochran set out to prove the usefulness of women pilots. Her own personal determination and unfailing faith in women aviators helped make the WASP program a great success. According to former WASP Madge Rutherford Martin, "Jackie had a dream, just like the rest of us. It took me years to understand she really wanted us to be able to fly everything just to show the men that it could be done. And, I think she largely succeeded in this."[64]

In 1944, Cochran had been saddened by the end of the WASP program, but did not dwell on the disbandment or the failure to attain military status. When the subject of military status resurfaced in the late 1970s, Cochran did not rejoin the fight. At a 1975 WASP reunion, former service pilot Teresa James tried to enlist Cochran's help in a new bid for military status. Cochran replied, "You'll never get anywhere in Washington. If I couldn't pull it off with all my contacts, it can't be done. I can't help you."[65] Of course, by that time Cochran was an elderly woman with a heart condition and only five years to live.

As for the WASPs, in 1977, Senator Barry Goldwater led the fight to recognize them as veterans. According to Goldwater, the only reason

8. The End of the WASPs

the WASPs were denied veteran status was because they were women. An elderly Cochran participated by issuing a few statements from her Indigo ranch. She pointed out that over 1,000 women pilots freed up an equal number of male pilots to serve in combat. "They [the WASPs] served their country," she told the national news, "they flew everything from B-29s down to Piper Cobras."[66]

Three decades after World War II, Congress recognized the WASPs as veterans of the Army Air Force.[67] Two years later, on March 8, 1977, they were finally granted military status. During the Congressional hearings, former Director Cochran did not even testify. At the time, her physical and mental health were failing. Instead, Bruce Arnold, son of the late commanding general, picked up his father's cause and was victorious.

Why did it take thirty years to militarize the WASPs? The answer lies in society's changing attitudes toward women. In 1944, women's issues were either dismissed or ignored, partly because Americans wanted to rebuild a stable society with sharply defined gender roles. Most women returned home and resumed their lives as housewives and mothers. Single women quickly married and began having children. While some women were unhappy, they quietly complied. In the 1970s, however, women launched a liberation movement and demanded fair and equal treatment. Suddenly, women's issues took on new importance, as women spoke out about injustices past and present. Therefore, militarizing the WASPs, which seemed unnecessary in 1944, was viewed as a just cause in 1977.

Recently, the WASPs have received some national attention. On May 10, 2010, Speaker of the House Nancy Pelosi presented former WASPs with the Congressional Gold Medal, the highest honor Congress can award to a civilian or civilian group. A total of 175 former WASPs attended the ceremony, along with over 2,000 representatives of deceased WASPs. Wearing her World War II-era Santiago blue uniform, former WASP Deanie Parrish accepted the award on behalf of her sister pilots. "Over 65 years ago," Parrish said, "we each served our country without any expectations of recognition or glory. We did it because our country needed us. All we ask is that our overlooked history will someday no longer be a missing chapter ... most of all in the history of America."[68] As of 2010, 300 WASPs were still living.

9

High Flying

The WASPs disbanded five months before the war in Europe ended. They were left, according to one WASP member, with "a bitter, unhappy feeling of having been given an important job to do, and then being rudely denied the privilege of finishing it."[1] Soon, other women would experience a similar disappointment.

When World War II finally ended, many women who had served the war effort felt a sense of loss. They had answered their country's call for help and were now no longer needed. Women had served as factory workers, government clerks and auxiliary military personnel in every branch of service. They were regarded as important contributing members of a nationwide effort to combat tyranny. Now, society expected them to return home and resume their lives as dutiful daughters and devoted wives and mothers.

Jacqueline Cochran, however, was unlike most American women. Before the war, she was far from the typical American housewife. Cochran was an American celebrity. She had made a career winning air races as a famous aviator and selling cosmetics as a successful businesswoman. During the war Cochran had garnered even more fame as director of the Women Air Service Pilots. When the war ended, she simply resumed her career and left other women to fend for themselves.

Cochran also had more freedom than most American women. Since she did not have children, Cochran did not have to worry about juggling motherhood with work. Nor did she have to contend with a traditional husband who expected her to stay at home. Cochran's husband, Floyd Odlum, always encouraged her to pursue her career. According to pilot Chuck Yeager, "Jackie was Floyd's hobby. He was damned proud of her, and really got a kick out of her."[2] Finally, Cochran had an economic advantage over most women. Her marriage to millionaire businessman Odlum gave her the financial freedom to pursue her dreams. According to Yeager, "Going for speed records cost a fortune, but he [Odlum] happily paid the bills."[3]

After the war, Cochran remained an important figure in aviation. As most American women returned to their homes, Cochran quickly resumed flying. She spent the next 15 years actively involved in aviation. In 1946, after the military had trained over a thousand women to fly military aircraft, Jacqueline Cochran was the only woman pilot who entered the Bendix Air Race. Flying a P-51B, she finished a disappointing second with a speed of 420.92 miles an hour and a time of four minutes and 52 seconds. Cochran also continued to set aviation records. In 1949, she set a new world record for propeller-driven airplanes, flying her P-51 Mustang at approximately 438 miles an hour.[4] Before her aviation career ended, Cochran would fly faster than the speed of sound, run for Congress and participate in America's new space program.

The post–World War II era was an exciting time in aviation. The war had led to advancement in aircraft. The propeller-driven airplane was taking a backseat to the new jet-powered engine. Once again, technology had revolutionized aviation. As a speed racer, Cochran could now go faster than she had ever dreamed.

The concept of a jet-propelled engine was recognized in the 18th century.[5] However, the lack of technology prevented its development and precluded any need for such an advancement. In 1910, Romanian aviation pioneer Henri Coanda built a bullet-nosed, two-wheeled thermo-jet powered biplane. In the next two decades, inventors attempted to build more modern jet aircraft. By the end of the 1930s, both the Germans and the British had invented jet engines.[6] Sir Frank Whittle of Great Britain and Hans J.P. von Ohain of Germany simultaneously, but independently, built similar jet airplanes, the Gloster Meteor and the Messerschmitt M-262. Whittle's engine was built by the British Thomson-Houston Company, an associate of the General Electric Company. Large-scale production of the jet engine airplane was only implemented toward the end of World War II.[7]

Germany invented the fastest jet plane, traveling at some 525 miles an hour. According to German Minister of Armaments and War Production Albert Speer, the Me-262 had "a fighting capability far superior to any plane the enemy had."[8] In 1944, America's War Department announced the use of jet-propelled fighter airplanes by the German Luftwaffe. At the time their value in war was debated. American engineers, however, predicted that the aircraft would "have speed, altitude and other performance characteristics beyond anything previously thought possible."[9]

Two years earlier, America had tested its first jet plane. On October 1, 1942, America's first jet aircraft, the Bell Airacomet (Bell XP-59A), took to the skies. Manufactured by the Bell Aircraft Company, its jet engine, the General Electric I-A, was modeled after the British design.[10] The Royal Air Force continued to test jet airplanes in both England and America. Meanwhile, the American Air Force began to develop other versions of jet aircraft.

In 1944, the American Air Force tested the Lockheed P-80 Shooting Star, also known as the F-80. Over a hundred Lockheeds were used in combat during World War II. The F-80 was followed by the Republic F-84 and North American F-86 Sabrejet.

A strong proponent of the jet airplane was American Air Force Commander General "Hap" Arnold. He was fascinated by Britain's first jet-propelled airplane, which he referred to as the "propellerless plane." According to Arnold, "I knew then and there I must get the plans and specifications of that jet plane back to the United States."[11] With Britain's permission, Arnold was able to bring the technology to America. In 1944, Arnold appointed a group of civilian scientists, under Dr. Theodore von Karman, to develop new weapons. The group designated the Scientific Advisory Group was encouraged to pursue any idea that was "not opposed to the law of science."[12] Even that rule was apparently flexible since Arnold suggested weapons of the "Buck Rogers" variety. Arnold instructed Dr. von Karman "to pay no attention to tomorrow's airplane, or the day-after-tomorrow's airplane, but to look into the future twenty years and determine what we would have to have then."[13] Arnold wanted a report that would guide him and future commanders of the United States Air Force. Jet technology revolutionized military aircraft.

Following World War II, research continued on jet-propelled aircraft. Soon the invention of jet engines would change civilian, as well as military, aircraft. In fact, the years between 1942 and 1952 were referred to as "the fastest 10 years in aviation history because of [the] almost unbelievable progress in jet propulsion."[14] America's jet age had arrived.

Like earlier aviation innovations, the jet airplane would need further testing. Following World War II there was no shortage of skilled male pilots ready to take up this challenge. America's first jet aircraft test pilot was Robert M. Stanley who flew the experimental Bell XP-59 A on October 1, 1942. Brig. General Laurence C. Craigie, however, was the first U.S. military pilot to fly a jet-powered airplane. He made a test flight the day after Stanley's historic flight.

Women test pilots, however, were rare. Between 1944 and 1953, only three women flew a jet airplane: WASP Ann Baumgartner; French pilot Jacqueline Auriol, who set a jet plane 100 kilometer speed record; and WASP leader Jacqueline Cochran. In 1944, Baumgartner was sent by Cochran on temporary air duty at Wright Field in Dayton, Ohio. There, she flew the YP-59, an experimental American jet. No other American woman would pilot a jet airplane until Jacqueline Cochran did so in 1953.

Like most aviators, Jacqueline Cochran was excited about the jet age. In 1948, she described jets as sensational aircraft destined to revolutionize civilian flying and air power in war. As she stated, jet airplanes "have opened new horizons in flying not yet scratched. Due to the turbo principal, the jet will develop jet weapons, capable of hundreds of miles per hour. The speed will be beyond any man's conception."[15]

Speed attracted Jacqueline Cochran. In 1948, she listed breaking the 2000-kilometer jet record as one of her greatest accomplishments. After the war, she committed herself to breaking two or three records in a jet airplane. In late 1948, she had made inquiries into flying jet planes. Despite her wealth, Cochran decided that buying a jet plane was impractical. After all, the plane could not be used in everyday transportation. Cochran said that owning a jet airplane would "have no practical value."[16]

Procuring a jet airplane was Cochran's first obstacle. Despite her celebrity status and close ties to Dwight Eisenhower, the United States government was unwilling to loan her a jet. She, therefore, contacted the Canadians. In December 1952, Cochran was hired as a part-time flight consultant for Canadair, Canada's airplane manufacturing company. In 1946, businessman John Jay Hopkins had bought Canadair for $10 million. By 1952, Canadair was a subsidiary of the American General Dynamics Corporation, with Hopkins as chairman and managing director. Floyd Odlum acquired a controlling interest in the company to help make Cochran's dreams a reality. Between 1950 and 1958, the company built 1,815 F-86 Sabre jets.

In the spring of 1953, Cochran wrote Hopkins and asked to rent an Orenda-powered F-86. Cochran willingly offered $10,000 to insure the aircraft. She also agreed to pay her own liability insurance, which amounted to $10,000 per flight. Cochran did not promise to break any records, expressing only hope that she would set a new speed record. As Cochran poetically wrote, "[I] hope to pluck that flower from the air as I dive through the sound barrier at Edwards, a desert of lake beds and salt flats perfect for scraping to a halt from fifty thousand feet."[17]

Cochran understood that jets were the present and future of aircraft, and to remain on top in the world of aviation, she knew she must master this new machine. Cochran was driven by raw determination. As Chuck Yeager commented, "When Jackie Cochran set her mind to do something, she was a damned Sherman tank at full steam."[18]

Cochran also recognized that the problems of flying a jet airplane were unlike any she had ever known. The only experience Cochran had in a jet aircraft was as a passenger. During a visit to France, she had flown in a jet fighter piloted by test pilot General Albert Boyd. Their speed reached over 600 miles per hour, the fastest Cochran had ever flown. Despite her independent spirit and mule-like pride, she enlisted help in flying jets. Cochran contacted Chuck Yeager, whom she considered America's best pilot.

In the early 1950s Yeager was well known in the field of aviation. He had built an impressive record as a flying ace of the Army Air Corps in World War II. During the war, he brought down one of Germany's new jet-powered Me-262, while flying a propeller-powered P-51. In 1946, he completed the first flight test pilot program. The next year, Yeager was the first pilot to break the sound barrier. In the 1950s, Yeager tested experimental jet aircraft at Edwards Air Force Base.

Cochran and Yeager had known each other for six years. They built a professional relationship, based on mutual admiration. Yeager was a frequent guest at the Cochran ranch and would often accompany

9. High Flying

Cochran on flights. Despite their closeness, the relationship was purely platonic, and neither Cochran's husband nor Yeager's wife, Glennis, felt threatened by the relationship. Cochran, after all, was known as a "man's woman" who felt more comfortable with male friends than female friends. Through aviation, Cochran and Yeager bonded. According to Yeager, "Jackie played a big role in my life, and I in hers."[19]

In 1953, Yeager took on the role of teacher and mentor to Jacqueline Cochran. He taught her everything he knew about jet aircraft. Flying side by side, Cochran and Yeager made several test flights before attempting to break records. During one of the runs, Cochran flew over a neighbor's poultry farm, traumatizing 2,000 chickens. Several hundred chickens suffocated when they crushed together in panic. Cochran paid for the damages.[20]

Yeager admired Cochran's skills as a pilot. Over and over again, he referred to her as a good pilot. Yeager described Cochran as a fast learner who made excellent landings.[21] He recognized her experience and natural flying abilities. "She can fly practically anything," Yeager said, defending her desire to break jet records.[22] He never doubted her abilities.

After making several test runs, Cochran and Yeager climbed to 45,000 feet. They plunged their airplanes into steep dives toward Edward's Air Force base and broke the speed of sound. In those few seconds, Cochran became the first woman to break the sound barrier. She described the experience like flying inside an explosion.[23] Unfortunately, the ground crew failed to record Cochran's feat, forcing her to make a second run. Not only was the Cochran the first woman pilot to break the sound barrier, she also flew faster than the speed of sound twice in one day. According to Yeager, Cochran loved to brag that she and Yeager "were the first and probably last man and woman team to break Mach 1 together."[24] The next day, Cochran broke the speed record, with an average speed of over 652 miles an hour.

On July 19, the *New York Times* published a story on Cochran's newest aviation feat. Described as "one of the nation's leading fliers," she is praised for topping the speed of sound. According to the article, ground observers were bombarded with a "blast like shock wave generated by the breaking of the so-called sound barrier." A photograph shows Cochran in her flying suit, smiling broadly over her latest accomplishment. Cochran had surpassed the 1951 record of Colonel Fred Ascani who had set the 100-kilometer mark in an American-built Sabre jet.[25]

True to her contradictory nature, Cochran loved airplanes but detested the noxious fumes that went with these machines. On her jet flights, she carried a bottle of perfume to counteract the smell of kerosene and sweat. The scent apparently lingered long after Cochran had gone. According to Yeager, "Every airplane she flew in smelled like a French whorehouse."[26] And, at least for year, every male pilot who flew after her was reminded that Jacqueline Cochran had been there first.

Aviation, however, was not the only interest Cochran had during the 1950s. In 1956, she decided to use her celebrity status to run for Congress. Once again, Cochran defied the traditional roles of women. In the 1950s, women were expected to stay home, tending their husband and children. Only a handful of women challenged traditional jobs held by men. Cochran had proved she could hold her own in the male-dominated world of aviation. She had beaten men in the air, but could she do the same on the campaign trail?

In the 1950s, as well as today, Congress was a male-dominated body of lawmakers. The first woman elected to Congress was Jeannette Rankin. Montana resident Rankin won a seat in the House of Representatives in 1916 and again in 1942. In 1922, the first woman senator, 87-year-old Rebecca Latimer Felton, took office. She was appointed to the position to complete the term of her deceased husband, Senator Felton, who had served one day. Hattie Caraway, also appointed to fill the seat of her deceased husband, later won two six-year terms. She served in the House from 1913–1921, and in the Senate from 1921–1931. In 1948, Margaret Chase Smith, a former member of the House, was the first woman elected to the Senate without a prior appointment to Congress.[27]

When Cochran ran for her seat, 17 women held seats in the House of Representatives. Only one woman, Margaret Chase Smith, held a Senate seat. In 1956, 31 women (including Cochran) vied for a seat in Congress. If Cochran were to win the 1956 election, she would be only the third woman to win a Congressional seat representing California. Cochran would follow fellow Republicans Mae Ella Nolan, who represented California from 1923–1925, and Florence Prag Kelly, who represented the state from 1925–1937.

9. High Flying

Cochran possessed several important attributes needed for politics. First, her name was well known. Just three years earlier, she had made news for breaking the speed of sound. Then, in 1954, Cochran had published her autobiography, *The Stars at Noon*. The book gave her name recognition as well as publicized her humble origins and inspiring rags-to-riches story. Secondly, Cochran was an attractive woman who photographed well and made a good impression on television screens. Her interest in make-up and stylish clothing would aid her on the campaign trail. Thirdly, Cochran had several political connections. A longtime friend of President Dwight Eisenhower, Cochran had friends in high places. Fourth, Cochran understood the art of timing. In 1955, popular representative John Phillips declined to run for another term as representative of California's 29th district. His retirement gave Cochran the opening she needed. Finally, Cochran had access to millions, courtesy of her encouraging husband, Floyd Odlum. The Cochran campaign would never lack for funding. Before the campaign ended, Cochran would outspend every other Californian candidate running for Congress.

Her attributes, however, could and would be used against her. Cochran's lack of education and her obsession with appearance could lead her to be labeled an empty-headed glamour girl who could fly airplanes but do nothing else. In one newspaper article, she was referred to as the "glamorous millionaire-aviatrix-cold cream manufacturer."[28] Her rich, influential friends made Cochran look like a greedy Republican, unconcerned about the poor. Floyd Odlum worried that even *he* might be a liability. He feared that Cochran's enemies would make an issue of their relationship, formed before Odlum was officially divorced.[29] Cochran's autobiography was also used against her. A newspaper pointed out that Cochran's "own story [painted her as an] immensely ambitious, ruthless individual ... using powerful friends to break the rules, using any weapon available to get what her unstable, immature but powerful ego wants."[30]

Backed by Odlum's money, Cochran flew around the state campaigning for the 1956 Republican nomination for the House of Representatives. She flew over 30,000 miles with a huge portrait of herself emblazed across her airplane. In one day alone, candidate Cochran flew over 540 miles and made speeches in several Californian towns.[31] If elected to Congress, Cochran promised to set up community-advisory committees to serve as liaisons between her and the people. "I'll never be more than six hours away from you by air," she promised one California community.[32] Cochran promised protection of existing water rights, supported the small businessman and organized labor within limits and rallied for strong military and foreign air bases.[33]

Cochran's primary opponent was Fred Eldridge, a rancher and former president of the Riverside County Farm Bureau. The two waged a nasty campaign. Cochran, capitalizing on the Red Scare, portrayed herself as staunchly anti–Communist and accused Eldridge of belonging to a subversive organization. Eldridge had been a member of the United World Federalists (UWF), an organization which advocated a one-world government. Cochran also labeled her opponent a "Red" for dedicating his 1946 book, *Wrath in Burma*, to the Chinese soldier and for his claim that American soldiers looted Germany. Meanwhile, Eldridge played on the Protestant majority's fear of Catholicism. He reminded voters of Cochran's Catholic faith and claimed that Catholic priests made immoral visits to her California ranch.[34]

When Cochran won the June primary, Eldridge chose not to endorse her in the fall campaign. The two exchanged terse letters in which Cochran reminded Eldridge of his lack of patriotism, and Eldridge expressed his desire that Cochran lose the November election. Eldridge accused Cochran of being a wolf in sheep's clothing and traveling "at the speed of sound with the voice of the people trailing behind like a sonic boom." He ended his letter with something of a curse. "Come November," Eldridge wrote to Cochran, "I hope your hide is tanned, your face made-up like a loser's — with egg and mud." Eldridge would not be disappointed.[35]

Following her primary victory, Cochran launched into the fall campaign with the same vigor she used to beat Eldridge. As the critics voiced, Cochran proved herself ambitious and ruthless. Throughout the campaign, Cochran used racial slurs to vilify her Indian-born opponent, Democrat Dalip Singh Saund. She described him as a "man from the lusty Punjab [whose] dark Hindu blood will cripple him in the Congressional cloakman."[36] She also suggested that her status as a Republican and close association with President Eisenhower would aid the district.[37]

9. High Flying

In one campaign flyer, Cochran is pictured with President Eisenhower, Vice-President Richard Nixon and several other Republicans running for office in California. The poster urged Californians to "join the team for America," to continue prosperity and vote Republican.

The November election was a close race. Cochran was unsuccessful at painting Saund as the boogeyman. Despite Saund's foreign birth and Hindu religion, he won the election, becoming the first Asian-born member of the House of the Representatives. When the ballots were counted, Cochran lost by only 3,000 votes out of 115,000 cast.

Why did Cochran lose the election? First, she did not involve herself in the community as much as her opponent did. According to one newspaper, "In all her [Cochran's] twenty years in the district, Mrs. Odlum has never served on a school board or a city council. She knows the President. She knows more generals and admirals than you can shake a stick at ... but as a representative in Congress, Mrs. Jacqueline Cochran looks very risky indeed."[38] Prior to running for office, Saund was a well-known successful farmer, member of the Lion's Club and Justice of the Peace. Meanwhile, Cochran was a glamorous celebrity who seemed out of touch with the people. Some of the neighbors felt Cochran was insincere and only became friendly when she wanted their votes. Others suggested that Cochran's alienation of Fred Eldridge cost her the election.[39] Most likely, Cochran's overbearing personality cost her some votes. Had she mended fences with Eldridge, rather than sending him a caustic letter after her primary victory, he might have helped with her campaign. As it turned out, she lost his vote and some the votes of his followers. Even Eldridge's campaign manager openly supported Saund in the November campaign.

Some 20 years later, Cochran remained bitter about the defeat. In a 1974 interview, she was still vexed that she was beaten by a Hindu, a man she described as a "damned Communist" and illegal immigrant at the time of the election. (Saund earned U.S. citizenship in 1949.) Cochran also regretted she did not run as a Democrat because then-Senator Lyndon Johnson offered to help her in the campaign. However, he was unable to support a Republican. Cochran remained a Republican partially out of loyalty to President Eisenhower.[40]

The 1956 Congressional race was Cochran's first and last experience with politics. She tried to wrangle an ambassadorship in the 1960s but was unsuccessful. Her first love remained flying, a passion which she had continued during the 1956 campaign. After the election, she set her sights on breaking new speed records. Cochran had reveled in the jet age. She had become the first woman to break the sound barrier and established a new speed record. If technology continued to evolve aircraft, Jacqueline Cochran would continue to conquer whatever frontier was next. The world of aviation was about to take another dramatic turn. In the late 1950s, Cochran turned her attention to America's new passion — space.

In 1957, the Soviet Union launched the first satellite into space. Sputnik I circled in the sky, frightening Americans and their government. The Soviets followed their success by putting a dog in space; and a Soviet man soon followed. In 1961, Soviet cosmonaut Yuri Gagarin became the first man to orbit the earth. America felt dwarfed by the Soviet's new accomplishments in space.

The creation of Sputnik brought several changes to America. In 1958, President Dwight Eisenhower signed the National Air and Space Act. This act led to the creation of the National Air and Space Administration. Science and math were emphasized in schools, with the hopes of producing future astronauts. The government poured money and rhetoric into America's new space program. In 1959, Americans were introduced to the Mercury 7, America's first astronauts. Two years later, newly elected President John F. Kennedy vowed that America would send a man to the moon by the end of the decade. "Space is open to us now," Kennedy explained, "and may hold the key to our future here on earth."[41]

Space may have opened for America — but not for American women. In the 1960s, NASA required that all astronauts be graduates of military jet test piloting programs and have degrees in engineering. At the time, no American women could meet those qualifications — not even America's most celebrity aviatrix, Jacqueline Cochran.

America's space program was bittersweet for Cochran. She applauded the technology but cursed her inability to participate. While the government courted talented pilots to join the space program, Cochran, a woman in her fifties, would never be considered. Even if she had been younger, Cochran could not meet NASA qualifications. She even took the same physical and psychological tests required of astronauts, but could not pass the tests.[42] Despite all of her accomplishments

9. High Flying

in the field of aviation, becoming an astronaut was out of Cochran's reach.

Yet, as late as 1961, Cochran had not given up her dream of going to space. "I wager that before I hang up my helmet," she told the Pacific *Stars and Stripes*, "I'll be going into space too. I may not be the first [woman] but after living through two eras of aviation, I can't imagine that the space age will pass me by."[43] She even expressed a desire to ride in a Mercury space capsule. "Just give me the opportunity," Cochran told one newspaper, after breaking a speed record in 1961.[44] Unfortunately, Cochran was born too late to become America's first woman in space.

The question remained: would Cochran do anything to help eligible women become astronauts? After all, she had a history of advancing herself and leaving other women behind. At the very least, Cochran could lend her money and her name to advancing women in space. But would she help others achieve what was closed to herself?

In 1961, Jacqueline Cochran was sworn in by NASA administrator James E. Webb as a special consultant to the air and space administration. Cochran, now with snow-white hair and a middle-aged figure, was no longer the glamorous girl pilot of the 1930s and 40s. Dressed in a white suit with a string of respectable pearls, she looked every bit the formidable person she had always been.

As a NASA consultant, Cochran set out to prove that women had the right stuff to be astronauts. She supported a private-funded program to test women for the space program. Cochran donated some $20,000 to the project. As with her work with the WASP program, Cochran actively recruited women for testing. According to the official qualifications, the women trainees had to be under 35 years of age and have 1200 to 1500 flying hours, rated on multi-engine airplanes and instruments. The program was actually the brainchild of Dr. Robert Lovelace, of the Lovelace Foundation for Medical Education and Research, who had previously worked with Cochran on high-altitude flight testing, and Brigadier General Donald Flickinger, who had designed the space flight tests for astronaut candidates. The Lovelace clinic administered the tests in Albuquerque, New Mexico.

Some 20 women applied to the program, and 13 passed. These

women included Geraldine "Jerrie" Cobb, a pilot who, like Cochran, had flown at the speed of sound. Cobb had been tested in 1960 (at the same time as Cochran) and was considered to be the top woman trainee. She successful completed 75 physical tests as well as psychological and psychiatric examinations. In 1961, she completed a two-week series of stress tests at the U.S. Navy School of Aviation Medicine in Pensacola, Florida. The other graduates, who completed Lovelace's women in space program in 1961, were Jean F. Hixon, Myrtle T. Cagle, Bernice Trimble Steadman, Iren Leverton, Geraldine Slone, Sara Gorlick, Gene Nora Stumbough, twin sisters Jan and Marion Dietrich, Mary Wallace Funk, Rhea Hurrle Allison and Jane Hart, the wife of Senator Phillip Hart of Michigan. The tests concluded that physically, women were as capable of going into space as men.

In 1961, Cochran spoke positively about women becoming astronauts and saw no barriers for women astronauts. After all, she explained, "Women can drive cars and planes as well as men do." She pointed out that women's participation in World War II proved they would just as capable as men to handle stressful situations. "The Battle of Britain," she said, "proved women can withstand the shock of war better than men can." At the time, the Soviet Union was training women astronauts, and Cochran did not want America left behind. "It shouldn't take that long for an American woman to leap into space. Certainly not if another country does it first."[45]

In 1962, Congress held hearings on the feasibility and fairness of including women astronauts. True to her contradictory nature, Cochran's testimony both supported and discouraged women astronauts. She told the committee that "physically women would make fine astronauts."[46] However, she also testified that while women airplane pilots are as emotionally and psychologically as fit as male pilots she was unsure as to how women would react in space: "I think there is no doubt women can go into space and be as successful as men, but I say I don't want to see it done in a haphazard manner."[47]

In her testimony Cochran said she wasn't against women in space. She simply believed the time was not right. She even stressed that she would not have used her own money to test women if she had not believed women should be astronauts. Even though Cochran's testimony was not the deciding factor on the matter, having America's most famous female flyer express reluctance did not help the cause of women astronauts.

Throughout the 1960s, Cochran continued to comment about women in space but would not fully endorse the idea. Even in 1968, a year before the moon landing, Cochran said, that "the space program is still in too early a stage to take on women pilots."[48] Why was Cochran reluctant to support her sister pilots in space? After all, she never doubted a woman's ability to pilot a space ship. According to Cochran, "There isn't one of us [experienced women pilots] who couldn't handle space craft." Nor did Cochran doubt women's physical strength, physiology, or mental capacity. "Physically," Cochran continued, "there's no problem either. Woman can take anything men can."[49] But, despite her praise of women's abilities, Cochran never actively crusaded for women to become astronauts. Sensing that her aviation career was coming to an end, Cochran was perhaps unwilling to pass the torch to others.

Cochran also believed in the gradual approach to change. During World War II, she constantly stressed that women were simply aiding, rather than replacing, the male pilots. She was content to give women pilots a small role in the war effort. She did not believe women should serve in combat and would not have approved of today's female military pilots. Cochran was not interested in remaking the social order, at least not for others. In her Congressional testimony, she suggested that women be trained as crewmembers first and then, maybe later as astronauts. In a letter to Jerri Cobb, Cochran urged her not to make waves over the cause of women in space. "Women," Cochran wrote, "for one reason or another have always come into each phase of aviation a little behind their brothers. They should, I believe, accept this delay and not get into the hair of the public authority about it. Their time will come and pushing too hard just now could possibly retard rather than speed that date."[50]

Cochran may have been trying to carve out a role for herself. In her 1962 Congressional testimony, she stressed that more tests were needed before sending women into space. She suggested the government choose 150 to 200 women, both pilots and non-pilots, and test their abilities on the ground. Cochran stressed, however, that such a program must not inhibit the current progress of the space program.[51] Perhaps she envisioned a female astronaut training program on the order of the Women's Air Service Pilots with herself at the helm. If she could not be America's first female astronaut, perhaps she could be responsible for another's success.

Despite endorsing a space training program for women, Cochran also cautioned against it. In fact, she feared such a program might interfere with the current space race.[52] Cochran believed it was economically unsound to train women. She foresaw NASA using millions of dollars to train women astronauts only to have them get married and/or become pregnant and drop out of the program. Ironically, Cochran never allowed her status as a married woman to interfere with her own goals. Yet, she felt other women who married might be a liability.

Cochran never fulfilled her dream of going to space nor did she create any training program. Now physically and mentally frail, Cochran retired from aviation long before women trained for space missions. The closest Cochran ever came to space was her attendance at spacecraft launchings. America's first woman in space, Sally Ride, made her ascent in 1984, four years after Cochran's death. Had Cochran been alive, she would have publicly praised Dr. Ride while privately seething her discontent at not being in her place.

The last 20 years of Cochran's life were filled with triumphs and tragedies. Throughout the 1960s, she continued to set speed records. In 1961, Cochran broke the speed of sound for a second time. In the same year, she set two new world speed records. In May 1963, she flew a 15–25 kilometer straightway course at 1,273.10 mph. A month later, flying a Lockheed F-104G Starfighter, Cochran flew a 100-kilometer closed course at 1,203.94 mph. Her record surpassed the speed record of French pilot Jacqueline Auriel, who had flown at 1,148 mph.[53] Cochran also set eight major speed records, flying a Northrop T-38. Two years later, she flew at Mach 2 — twice the speed of sound.[54] In 1964, Cochran set three new speed records in a Lockheed jet Starfighter. In one run, she flew 1,429 mph, the fastest air speed ever flown by a woman. By the end of the decade, Cochran held over 100 air records.

In 1963, she sold Jacqueline Cochran Cosmetics. Her successful business had a part of her life for thirty years. In the 1950s, the Associated Press voted her Business Woman of the Year, two years in a row. Cochran

9. High Flying

was almost as well known for her Flowing Velvet facial cream as she was for her aviation feats. Jacqueline Cochran owned much of who she was to the beauty business. It had taken her out of poverty and helped her become a glamorous beauty. Cochran's desire to promote her cosmetics business led to her decision to take up flying. Her company's slogan "Wings to Beauty" reflected her two greatest passions.

After selling the company, Jackie and Floyd settled down into a quiet life at their Indigo ranch. They continued their associations with some of the world's most powerful people. She and Floyd were frequent guests in the Johnson White House. Cochran enjoyed a warm friendship with Lyndon Johnson, whom she had met at the 1937 Robert J. Collier Trophy award ceremony. Cochran continued to accept awards and honors, and was often a guest speaker. In 1971, she was admitted to the Society of Experimental Test Pilots. In 1975, the Air Force Academy presented Cochran with the Sabre Plaque award. Throughout her lifetime, she was given over 200 awards and trophies. Cochran also continued to follow the progress of space exploration.

The last fifteen years of Cochran's life were filled with many heartaches and disappointments. In the late 1960s, the Odlums experienced their first financial hardships as a couple. Floyd's son, Bruce Odlum (the son of first wife Hortense) asked his father to finance a real estate venture. The business venture failed and the Odlums lost an enormous amount of money. Although not financially ruined, Floyd and Jackie were forced to sell their beloved 340-acre ranch, a decision which deeply hurt Cochran. Later, Bruce Odlum committed suicide. His death was a tragedy for both Floyd and Jackie.

The 1970s was a particularly tragic decade for Jacqueline Cochran. After nearly 40 years of aviation triumphs, her frail body could no longer keep up with her adventurous spirit. A botched appendectomy at age 20 had caused Cochran pain most of her life. Now, doctors installed a pacemaker to keep her frail heart beating. Cochran's heart could no longer withstand the shock of high-speed flying, and in 1971 she was forced to give up flying altogether. Losing her wings was a difficult adjustment for Cochran. She had always tried to fly at least 40 hours a month and would look for any excuse to take to the air. "I was only happy in an airplane, completely," Cochran commented in 1974.[55] Throughout her aviation

Jacqueline Cochran

career, she had logged some 15,000 flying hours in both propeller and jet airplanes.

Cochran spent the last five years of her life in Indigo, California, hosting dinner parties, attending WASP reunions and caring for her ailing husband, Floyd. Five years after giving up flying, Cochran faced another tragedy when her beloved Floyd died. Diagnosed with rheumatoid arthritis in the 1940s, Floyd Odlum was confined to a wheelchair during the last years of his life. Throughout their 40-year marriage, he remained Cochran's greatest supporter, both financially and emotionally. Odlum often feared that Cochran was not taken seriously because she was a woman with a rich husband who could finance her every dream. "[Some] insinuate," Odlum once wrote to his wife, "that I have been the pilot of your life everywhere but in a plane. If only they knew how it really is: I was made from *your* rib."[56] Odlum's death took the life out of Jacqueline Cochran. Four years later, she followed him to the grave. Cochran had requested a simple funeral, an irony for a woman who had spent her whole life being noticed. In accordance with her wishes, she was laid to rest in a plain pine coffin, adorned with yellow roses.

* * **

At a cemetery in Indio, California, Jacqueline Cochran quietly rests as a largely unknown pioneer of aviation. Her simple but unique gravestone is emblazed with an F-104 Starfighter, one of several jet airplanes she flew to fame. The stone lists her birth as 1906, although she was never really sure the year she was born. On the stone, she is recognized as a colonel in the United States Air Force Reserve. Her two greatest accomplishments are also listed: Leader of the Women's Air Force Service Pilots and the first woman to fly faster than the speed of sound.

Jacqueline Cochran lived an extraordinary life. From the 1930s to the 1970s, she accomplished great feats in every decade. In the 1930s, she earned a pilot's license and became the first woman to win the prestigious Bendix Air Race. When World War II began, she flew a bomber across the Atlantic Ocean and then led American women pilots to England to aid the Royal Air Forces' auxiliary branch. After America entered the war, she formed the Women's Air Service Pilots, giving female aviators a chance to fly military aircraft and aid the war effort. In the

9. High Flying

1950s, she flew faster than the speed of sound. A decade later, she flew twice as fast and set new speed records. Although Cochran was forced to give up flying in the 1970s, she was still recognized for her flying skills. In 1971, Cochran was enshrined in the Aviation Hall of Fame. The same year, Cochran retired as a colonel in the Air Force Reserve after receiving the Legion of Merit award.

Throughout her life, Jacqueline Cochran had two goals. Her first goal was self promotion. As a poor child in a Florida mill town, Cochran was determined to live a better life. She became a popular beautician and profitable businesswoman. She earned a pilot's license and set out to be a famous aviator. As a pilot, she desired to win races and set speed records not to promote women aviators but to promote herself. Her 1956 bid for Congress was another attempt at self promotion. She was more interested in speaking her mind than helping her California neighbors or advancing women in politics. When Cochran was unable to join the space program, she made some attempt to promote women as astronauts but did not fight for their participation. If Jacqueline Cochran could not be the first woman in space, she would not crusade for any other woman to take her rightful place.

Cochran's second goal was the advancement of aviation. A born flier, she pushed airplanes to their limits to see how fast she could go. Cochran was even willing to sacrifice the cause of women in aviation, for the good of aviation. Over and over again, she let herself be used as a woman pilot to show how easy it was to pilot an airplane. In the 1930s, she successfully flew the Gee Bee Racer, an airplane so deadly it was called the widowmaker. In 1941, she flew a bomber across the Atlantic partly to shame male pilots into aiding the Allied war effort. Cochran also tested the P-47 to prove to other pilots that if a woman could fly the airplane, anybody could.

* * **

Jacqueline Cochran was not an educated woman but, she had a sense of history. She often wondered how history might portray her. She wondered who would go through her letters, memorabilia and other personal belongings. And she wondered what that person might conclude of the woman and pilot known as Jacqueline Cochran. "From this treasure and dross they tell me a scholar can re-weave a life." Perhaps, through history she saw herself living again. She predicted a female biographer — a "beautiful stranger in a tailored suit who'll wear lipstick and comb her hair at coffee breaks. [But will she] appreciate my tatting as well my flying skill?" She added, though, that no biographer could capture the thrill she found in flying.[57] Nor can one really capture Jacqueline Cochran — a true woman of contradictions.

In 1954 Cochran posed an interesting questions when she asked, "Why does a women with a fine family and a cosmetics business go on risking her life for [aviation]?" "For an answer," Cochran wrote, "you have to climb 50,000 feet on a bright summer day. Up, up, up you go until the sky is deep blue like the vesper hour of the evening. For a fleeting moment, you're the last person on earth. Then you glance up and — look! Thousands of stars are winking down at you, at noon on a sunshiny day! They you know you're not alone at all. There's someone up there with you, peeping over your shoulder at the control panel. That's where I found my answer. To sit for a moment on the doorstep of God's Heaven, isn't it worth the risk?"[58]

10

Postscript

Jacqueline Cochran, America's premier female aviator, had the misfortune of living too long. Had she died in some horrific plane crash or disappeared like her friend Amelia Earhart, Cochran's name would be legendary. Instead, she lived to an old age and died quietly of natural causes. By then, Cochran's youth, beauty and celebrity had faded. Despite her many accomplishments, she remains a shadowy figure in American history.

Unjustly, it is Earhart, rather than Cochran, whose name is immediately associated with women in aviation. Earhart is the subject of countless biographies, as well as conspiratorial books and television specials (some reputable and some not) investigating her mysterious disappearance. Recently, she has become a character in one movie and the main subject in another. In the sequel to *Night at the Museum*, titled *Battle of the Smithsonian*, Earhart (played by Amy Adams) is portrayed as a spunky redhead with a bad sense of direction and eyes for the main character, played by Ben Stiller. Much of the movie takes place in the Air and Space Museum. While, the Tuskegee Airmen are given a cameo appearance, Jacqueline Cochran is absent. A 2009 film, entitled *Amelia*, offers a more realistic portrayal of Earhart, but still ignores Cochran. Earhart, played by Hilary Swank, is portrayed as a gutsy, determined pioneer of aviation and an outspoken feminist. Yet her friendship with Jacqueline Cochran is never mentioned, much less the fact that Cochran was a much more skilled flyer than Earhart. The movie, which places great emphasis on Earhart's last flight, makes no mention of the voyage being planned at Cochran's California ranch or Cochran's attempts (psychic or otherwise) to find her. In the film Earhart is often hailed as "America's Number-One Female Flier," a title that, in actuality, belongs to Jacqueline Cochran.

Jacqueline Cochran

In 1997, Leslie Haynsworth and David Toomey published a book about women in aviation. The book's title is *Amelia Earhart's Daughters,* implying that all women pilots somehow evolved from Amelia. While several pages are devoted to Cochran's career as an aviator, it is Earhart's name which adorns the cover. After all, placing Amelia Earhart's name on a book results in instantaneous recognition and immediate association with aviation. Putting Jacqueline Cochran's name on a book cover does not convey the same meaning.

Recently, Cochran has received some attention from writers and historians. Prior to the 21st century, the only work on Cochran was a children's book by Marquita O. Fisher entitled *Jacqueline Cochran: First Lady of Flight.* Published in 1973, the book chronicles Cochran's life from the "girl who didn't belong" to the woman who broke the speed of sound. In 2001, writer Enid Shomer published a brief book of poems dedicated to Cochran. The book, entitled *Stars at Noon: Poems from the Life of Jacqueline Cochran,* includes stories and letters about Cochran which are crafted into poetry. The book's intention is to exemplify Cochran's determination and strength. In 2002, Global Science Productions produced a 50-minute documentary on Cochran entitled, *Jackie Cochran: First Lady of Flight.* The video explores Cochran's life from her humble Florida roots to her rise as an aviator to her death. The documentary begins with fellow aviator Chuck Yeager describing Cochran's gutsy, take-charge personality. Included in the video are some scenes from a 1967 interview with Cochran.

In 2000, Margo McLoone published *Women Explorers of the Air,* a small book containing easy-to-read biographies about women in aviation. Cochran is profiled along with Harriet Quimby, Bessie Coleman, Amelia Earhart and Beryl Markham. Cochran is also discussed in Deborah G. Douglas's 2004 work *American Women and Fight Since 1940.* The book discusses Cochran's aviation career, her work with the WASP program and her views on women in the space program.

The most recent work on Cochran is a 2007 biography written by journalist Doris Rich. *Jacqueline Cochran: Pilot in the Fast Lane* chronicles Cochran's aviation career as well as her interest in politics. Rich includes a chapter on Cochran's 1956 unsuccessful run for California's 29th Congressional District, a race partially lost by Cochran's caustic personality.

10. Postscript

Cochran's political views are also discussed. Despite being a friend of moderate President Dwight Eisenhower and liberal-minded President Lyndon Johnson, Cochran was a staunch conservative, who moved further right as she grew older.

Although Cochran is not as well known as Amelia Earhart, she remains part of aviation history. Her records are proudly displayed at the Aviation Hall of Fame in Dayton, Ohio. At the Mighty Eight Museum in Savannah, Georgia, she is honored in an exhibit for her contributions as director of the Women's Air Service Pilots. At the National Air and Space Museum in Washington D.C., she is recognized as a pioneer of aviation. Her 1938 Bendix Trophy is part of the Smithsonian Collection. Since 1967, a portrait of Jacqueline Cochran hangs in the Smithsonian Art Museum. Painted sometime around 1950, the picture — a ghostly image of Cochran in a pale pink formal gown — portrays her glamorous side. At the United States Air Force Academy in Colorado stands a permanent exhibit of Jacqueline Cochran memorabilia. Included in the collection is her blue Air Force uniform, some of her awards and a large portrait of Cochran in an orange flying suit. The WASPs Museum in Sweetwater, Texas, has recently opened a new Jacqueline Cochran exhibit.

Since her death, Cochran's name has appeared on stamps, buildings and public events. On a 1996 airmail postage stamp, she is honored as "Jacqueline Cochran: Pioneer Pilot." She is pictured wearing a flying helmet and goggles with a symbol of the Bendix Air Race in the background. In Thermal, California, a regional airport she often used was renamed for her. Since 2002, the airport has hosted the Jacqueline Cochran Air Show, also named in her honor.

Jacqueline Cochran, the true heroine of aviation, was one of most successful American women of the 20th century. According to the WASP Museum, she "went higher and faster into the frontiers of aviation history then any woman before ... [blazing] a trail for other heroic women to follow." She may not have actively crusaded for women but, through her own accomplishments, she inspired and aided others. Using an aviation-like analysis, Cochran left the secret of her success. "If you will open your power plants of vitality and energy, clean up your spark plugs of ambition and desires, and pour in the fuel work, you will be likely to go places and do things."

Appendix One

Biographical Sketches

General Henry Harley Arnold, USAF, General of the Air Force, 1886–1950

General Henry Harley "Hap" Arnold, was born on June 25, 1886, in Gladwyne, Pennsylvania. After graduating from high school, he passed the examination for an appointment to the U.S. Military Academy at West Point, New York. He entered West Point in 1903, and graduated with the Class of 1907. Commissioned a second lieutenant in the Infantry, the Army assigned him to the Philippine Island of Luzon, where he conducted a survey of the island. In 1910 he was sent to Governor's Island in New York City where he took his first test spin in a Curtiss biplane with famed aviator Harry N. Atwood. Lieutenant Arnold requested and received assignment to the Army's aviation branch, then attached to the with the Signal Corps. Passing the pre-flight instruction physical, he reported to the Army's Flying School, then located at Dayton airfield, which was operated by Orville and Wilbur Wright. After completion of a two-month course, during which he mastered the controls of a two-seater Wright biplane (that had a top speed of 40 mph), Lieutenant Arnold reported to the Army Flying Field at College Park, Maryland. There, on July 6, 1911, Arnold received his pilot's certificate and designation. From the start of his career as an aviator, he established several military aviation records, first of which was being the only man ever to ascend to 3,260 feet. This was accomplished on the same day he received his wings as an Army aviator. In April 1912, Arnold had successfully attained an altitude of 2,500 feet, made a dead-stick landing within 150 feet of a designated point, conducted a military reconnaissance of 20 miles cross-country at an average altitude of 1,500 feet, and carried a

Appendix One — Biographical Sketches

passenger while in flight. Lieutenant Arnold was also the first pilot to fly in an airplane — a Wright Type C, the first cockpit-type plane to be purchased by the Signal Corps. For the next two years, when Arnold wasn't flying he was teaching new aviators at College Park. In October 1912, Arnold and several other officers reported to Fort Sill, Oklahoma, where he and two other officers conducted an experiment by directing artillery fire from the air. At Fort Sill, Arnold became the first officer to use a radio while in flight for reporting observations. Reassigned to the Officer of the Chief Signal, Arnold made his first contact with a fellow aviator on the General Staff— Captain (later Brigadier General) William E. ("Billy") Mitchell. By the time of his re-assignment to the Infantry in April 1913, Arnold had distinguished himself as an Army aviator. In March 1916, after a brief tour of duty in the Philippines, Arnold was again re-assigned, this time to the Army's newly established Aviation School, located in San Diego, California. In February 1917, the War Department ordered Arnold to the Canal zone in order to organize and command the first aerial defense of that strategic area. With the U.S. entrance into World War I (April 6, 1917), Arnold was given command of a pursuit school located at Carlston, Florida, then assignment to Washington, D.C., as head of the Information Services of the Aviation Section with a temporary promotion to the rank of major (June 27, 1917); then colonel I (August 5, 1917). When the Bureau of Military Aeronautics was established in early 1918, the Army assigned Arnold as assistant executive, then executive, and finally assistant director. Arnold reached France in November 1918, too late to join in on the fighting.

During the immediate post–World War period (1919–1921), Arnold went on serve as district supervisor, Western District of the Air Service at Coronado, California, and served there until May 30, 1919. He then served as the Air Officer of the Ninth Corps Area in San Francisco until August 11, 1920, when he transferred to the newly established Army Air Service. From the outset, Arnold was one of several Army aviators who began pressing the Army to create a separate air force. In fact, after graduating from the Army's Industrial War College in 1925, Arnold reported to the Office of the Chief of Staff, where he reassumed his old position as Chief of the Information Services. During the court-martial of his friend and fellow aviator Brigadier General William "Billy" Mitchell, Arnold took the stand on behalf of the defense as he vigorously defended Mitchell's call for an independent air force. In 1926, Arnold returned to Fort Riley, Kansas, where he assumed command of the Air Corps troops stationed there. Later, as commanding officer of March Field in Riverside, California, Arnold, now a lieutenant colonel, organized the first round trip flight of ten of the Army's new Martin B-10 bombers, the direct descendant of the Army's B-17 "Flying Fortress," from Bolling Field, Washington, D.C., to Fairbanks, Alaska, and back. For this and other aviation-related pioneering feats, the Army awarded Arnold the Distinguished Flying Cross (February 1936). On February 11, 1935, he was promoted to the temporary rank of brigadier general. That same year he was also awarded a second Mackay Trophy for the Alaska flight. In September 1936, after the death of General Westover in a plane crash, Arnold became chief of the Army Air Corps. As Europe once again drifted toward war, General Arnold became a strident voice for preparedness and for the creation of a strong, separate air force.

When the United States entered World War II after December 7, 1941, Arnold became commanding general of the Army Air Force (March 1942). In October of that same year, Arnold received the Distinguished Service Medal. In March 1943, Arnold received his first Air Medal. As the bombers and fighters of the Army Air Forces bombed German and Japanese cities, and American fighters swarmed over Germany and Japanese-held territory, Arnold was promoted (temporarily) to the rank of general and later, on December 21, 1944, as a five-star general of the Army. From the outset of his career, General Arnold believed in loyalty, a trait he held in high regard, as evidenced by his dogged defense of Brigadier General Billy Mitchell in 1925.

General Arnold's career mirrored the growth and expansion of military aviation and, more importantly, the growth and expansion of Army Aviation that went from a force that consisted of two Wright Brothers airplanes to that of a force that consisted of five major air forces (the 8th, the 15th, the 10th, the 5th, and 20th), thousands of aircraft of all types, and hundreds of thousands of pilots, ground officers, and enlisted men and women, not to mention the thousands of bombers (B-17s, B-25s, B-24s), the B-29 "Superfortress," fighters, observation aircraft, and the first jet fighters. Indeed, it was General Arnold's Army Air Corps that would inaugurate the concept of strategic bombing and be the force that carried the United States military into the atomic era with the dropping of the first two atomic bombs on Hiroshima and Nagasaki, Japan. Most important, however, was General Arnold's call for, and the eventual realization of, an independent air force in 1947. Upon General Arnold retirement in February 1946, the United States had built perhaps the greatest air force in the world, a fact that remains as true today as it did at the end of World War II. A large portion of this is due to his Herculean efforts before, during and immediately after World War II.

In recognition of his outstanding achievements as an Army officer and aviator, as well as head of the Army Air Forces during World War II, Congress, on May 7, 1949, officially designated Arnold as "General of the Air Force" (81st Congress, Public Law 58). General Arnold died on January 18, 1950, and was buried with full military honors in Arlington National Cemetery.

Vincent Hugo Bendix, Automotive and Aviation Pioneer, 1882–1945

Vincent Hugo Bendix was born on August 12, 1882, in Moline, Illinois, the son of Swedish immigrants Jann Bengtsson and Anna Danielson. After receiving his early education in Moline and Chicago, Bendix moved to New York City, where he worked in the maintenance department of a local hospital. It was there that he became fascinated with electricity. After working as an accountant in a brewery and in the office of the Lackawanna Railroad, Bendix went on to study law with a New York law firm, although his strong interest in mechanical engineering and electricity led him to leave the practice of law and into the burgeoning field of automotive design and manufacture. In 1901, he was hired by Glenn Curtiss, the famed manufacturer of aircraft, to work on the Torpedo motorcycle. It was during his work with Curtiss that Bendix became very interested in the internal combustion engine and power-propelled vehicles. Leaving Curtiss's company, Bendix went to Chicago, Illinois, to work for the Holsman Company as general sales manager. Leaving the Holsman Corporation in 1907, Bendix formed his first company and manufactured the Bendix Auto Buggy. Although the business failed,

Bendix retained his interest in the auto industry, selling Cadillacs and later serving as sales manager of the Haynes Motor Company. It was in 1910 that Bendix first developed the idea of an electric starter for autos, replacing the hand-crank method used to start automobiles. He received his first patent later that same year for his invention. Bendix likewise went to work with the F. A. Ames Auto Buggy Corporation, located in Owensboro, Kentucky. While with the Ames Corporation, he formed a lifelong partnership with Herbert Sharlock. It was with Sharlock that he started the Bendix Corporation in 1911. After some resistance to his invention, Bendix's idea took off, and he successfully marketed the electrical starter. The first starters were placed in the 1914 Chevrolet "Baby Grand," of which some 5,500 drive units were sold. Throughout the ensuing decade (1914–1924), Bendix became one of the leaders in the auto industry, manufacturing such things as brakes and carburetors. Caught up in the "flying fever" of the decade, Bendix formed the Bendix Aviation Company in 1929. Later, in 1931, Bendix organized and sponsored the Bendix Transcontinental Air Race. Winners of the Bendix Trophy included Major James J. Doolitle, of the U.S. Army Air Forces (1931) and Jacqueline Cochran (1938). In 1942, Bendix resigned as chairman of the board of the Bendix Aviation Corporation. He went on to form the Bendix Helicopter Corporation in 1944, for which he had hoped to develop a four-passenger helicopter for both civilian and military use. Before his ideas on vertical flight could be realized, however, Victor H. Bendix died of a heart attack on March 27, 1945, at his home in Chicago, Illinois.

The world of travel owes much to the genius of Victor Bendix who was a pioneer and leader in both the aviation and automotive fields during the 1920s and 1930s. While possessing extravagant tastes, a factor which led to financial difficulties over the years, Bendix nonetheless is considered one of the giants of both the automotive and aviation industries.

Amelia Earhart, Aviatrix, 1897–1937

Amelia Earhart, the famed aviatrix, was born in Atchison, Kansas, on July 24, 1897. After graduating from high school, Earhart had planned to attend college but her plans changed upon the outbreak of America's

Appendix One — Biographical Sketches

entrance into World War I. During the war she worked as a military nurse in Canada, and at the war's end in 1918, became a social worker at the Denison House, located in Boston, Massachusetts. In early 1920, after watching the various stunt flyers and taking a plane ride for $1, she decided to take flying lessons. After ten hours of instruction and several near-fatal crashes, she earned her pilot's license. By 1922, she was flying a plane of her own. In 1928, in an airplane named *Friendship*, Amelia became the first woman to cross the Atlantic Ocean albeit as a passenger. Covering the story was publisher George Putnam, who later became her husband and unofficial biographer. Earhart's flight brought her much publicity over the next few years, and on May 20–21, 1932, Amelia Earhart became the first woman to solo across the Atlantic Ocean, winning acclaim on both sides of the Atlantic. In recognition of her achievement, President Herbert Hoover presented her with a medal. Several years later she successfully flew from Hawaii to California.

Still hoping to set an endurance record, Amelia Earhart set out to fly around the world in a twin-engine Lockheed Electra. She and her navigator, Fred Noonan, set out from Miami, Florida, and headed to the Sahara Desert, and from there to Thailand, Singapore, Java, and Australia. From Australia she and Noonan flew to Lae, New Guinea, for Howland Island near the Marianas Island of Saipan. It was there that the U.S. Coast Guard lost contact with Earhart. After receiving a message from her on July 2, 1937, the plane disappeared from site. After a frantic search, the Coast Guard declared Earhart and Noonan lost at sea somewhere over the Howland Islands. The U.S. Navy conducted an extensive search but never found even a trace of the wreckage or any bodies. Earhart's disappearance raised speculation that perhaps she had flow too close to the Japanese island of Saipan and had been captured and executed. Others have speculated that Earhart was possibly on a spy mission for President Franklin D. Roosevelt in monitoring the Japanese build up in the Pacific, though there has been no evidence that this was the case. Other theories hold that she simply ran out of fuel or had crashed into the ocean. In any event, none of these theories have ever been proven. In 1967, her husband, George Putnam, published a biography, entitled *Soaring Wings*, as a tribute to her pioneering work as a female aviator. During the early 21st century, there has been a renewed interest in Earhart and her career as an aviatrix as well as her attempt to circumnavigate the earth. Most important of this revival is the film entitled *Amelia*, which comes very close in depicting the woman behind the legend. Like Jacqueline Cochran, Earhart broke the traditional boundaries held by men in the aviation field. Indeed, it was to her credit, as well as Cochran's, that women became accepted in what had been a male-dominated field, and blazed a path that today is carried forth by military and civilian female pilots.

Colonel Oveta Culp Hobby, U.S. Army, Women's Army Auxiliary Corps (WAAC)— Women's Army Corps (WAC), 1905-1995

Colonel Oveta Culp Hobby was born in Killeen, Texas, on January 1905, the daughter of I. W. and Emma (Hoover) Culp. In February 1931, she married William P. Hobby. Her early education included attendance in the public schools in Killeen, Texas, and Mary Hardin-Baylor College, in Belton, Texas. Deeply interested in politics, she became a Parliamentarian for the Texas House of Representatives and later, Assistant City Attorney in Houston. In 1931, she married William P. Hobby, the former governor of Texas and publisher of the *Houston Post*. She remained in Houston with her husband, running the Post until 1941, when the United States entered World War II. Coming to the Nation's Capitol as a "$1 a year person," she accepted a position inside the War Department as head of the War Department's Women Interest Section. From 1942 to 1945 she served as director of the newly created Women's Army Auxiliary Corps (WAAC). On July 5, 1943, Director Hobby was sworn in as a colonel, thus becoming the first woman to be admitted to the new component of the U.S. Army. According to the provisions of the legislation, the Army had 90 days in which to dissolve the women's Auxiliary Army Corps, so that by September 30, 1943, all members of the Women's Army Auxiliary Corps had to be enlisted or commissioned in the Woman's Army Corps, or be discharged. While serving as director of the WAAC, Colonel Hobby designed and fielded a green and gold service ribbon for those members of the Women's Army Auxiliary Corps who elected to remain with the Women's Army Corps.

With the United States fighting a two-front war, there was a critical drain on manpower. Eventually, over 150,000 American women served in what would become known as the Women's Army Corps, or WAC. Both the Army and American public initially had some difficulty in accepting the concept of women in uniform though women, in fact, had served briefly during World War I (1917–1918). Both political and military leaders realized, however, that women could supply the additional resources so desperately needed in the military and industrial sectors. Given the opportunity to make a major contribution to the war effort, women served with distinction on all fronts. Members of the WAC were the first women (other than nurses) to serve within the ranks of the United States Army during World War II and beyond. Through the hard work and wise leadership of Colonel Oveta Culp Hobby, women soon found themselves playing a critical role in the war effort and in the achievement of final victory in 1945. As a result of her efforts and success as director of the Women Army Corps, Colonel Hobby was awarded the Distinguished Service Medal and the Philippine Merit Medal.

At the conclusion of World War II, Colonel Hobby returned to Texas to assist her husband in running the *Houston Post*, as well as a newly acquired television station. In 1953, President Dwight D. Eisenhower named her the head of the Federal Security Agency. Later that year, her position was elevated to a Cabinet post and renamed the Department of Health, Education, and Welfare, with Hobby becoming its first Secretary. In 1955, Hobby returned to Houston to take care of her ailing husband, who passed away in 1964. She succeeded her late husband as the editor and chairman of the board of the Houston Post Company in 1965. She also became director of Station KPRC and KPRC-TV. She likewise authored a parliamentary law textbook, *Mr. Chairman*, and became a syndicated columnist. She lived in Texas the remainder of her life and died in 1995 after a long life of extraordinary achievements.

Nancy Harkness Love, Aviatrix and founder of the Women's Auxiliary Ferry Squadron, 1914–1976

Nancy Harkness Love was born on February 14, 1914, in Houghton, Michigan. From an early age, Nancy Harkness developed what has been described as a strong interest in aviation. At age 16, she flew for the first time, and within a month had earned her pilot's license. Having attended some of the finest schools for young ladies of the time (Milton Academy in Massachusetts and Vassar in New York), she nonetheless was caught up in the aviation fever that had spent the nation during the 1920s.

Nancy Harkness's love of aviation soon attracted her to Robert Love, a reserve major in the U.S. Army Air Corps, whom she married in 1936. Together they formed their own aviation company — Inter City Aviation, for which Nancy served as a pilot. She also flew for both the Bureau of Air Commerce, and in 1937–1938, as a test pilot, performing safety tests on various aircraft modifications and innovations. She was the first woman aviator to test and land an airplane using a tricycle landing gear which, in time, became standard on most aircraft of the day. She also encouraged the idea that towns place their names on water towers, thus making them identifiable navigation aids for pilots.

With the outbreak of the Second World War in Europe, Love wrote to Lieutenant Colonel Robert Olds, who was then establishing a Ferrying command within the U.S. Army Air Corps. She informed him that she had 49 well-qualified female pilots available for service who could assist in ferrying aircraft from the factory to various Army Air Corps bases. Colonel Olds took the suggestion to General "Hap" Arnold, who rejected the proposal.

When her husband was called to active duty in 1942 as deputy chief of staff of the Ferry Command, Love accompanied him and soon acquired a position within the Air Transport Command. Commuting from her job in Baltimore, Maryland, in her own airplane to Washington, D.C., Nancy Love soon gained the attention of Colonel William Tunner, whom she convinced of the need to utilize qualified women pilots in the ferrying of aircraft. In response, Tunner asked Love to draw up plans for what later became the Women's Auxiliary Ferry Squadron, or WAFS. Love soon became the WAFs' commander, and in September 1942, the women pilots of the WAFS began ferrying aircraft out of the New Castle Army Air Field, located at Wilmington, Delaware, as part of the Air Transport Command's 2nd Ferrying Group. By June 1943, there were four different squadrons of WAFS operating at different points in the United States. In August 1943, the WAFS merged with the newly created WASP

(Women Airforce Service Pilots), headed by aviatrix Jacqueline Cochran. Nancy Harkness Love was made executive director of the WASP program. Between September 1942 and the groups disbandment on December 20, 1944, the WASP delivered some 12,650 aircraft of 77 different types to various Army Air forces bases nationwide. In recognition of her contribution to the war effort, the War Department awarded Nancy Harkness Love the Air Medal for her "operational leadership in the successful training and assignment of over 300 qualified women fliers in the flying of advanced military aircraft." At the end of the war, Nancy Harkness Love retained her interest in aviation and promoted the deeds rendered by both the WAFS and WASPs during World War II. She died on October 22, 1976.

General of the Army George Catlett Marshall, U.S. Army, 1880–1959

General of the Army George Catlett Marshall was born in Uniontown, Pennsylvania, on December 31, 1880. After receiving his early education in Uniontown, Marshall entered the Virginia Military Institute in 1897, and graduated in 1901. Commissioned a second lieutenant in the Army on February 2, 1901, he reported for duty with the 30th Infantry Regiment (1902–1903) then on duty in the Philippines, during the pacification campaign there. Leaving the Philippines, Marshall's next assignment was at Fort Reno (1904–1906), and later as an instructor at the Army War College (1908–1910). He returned to the Philippines, where he was stationed from 1913–1916. Returning to the United States in 1917, Marshall was assigned to the staff of the 1st Division, Chief of Operations, 1st U.S. Army, and Chief of Staff, 8th Army Corps. During combat operations in World War I, Marshall participated in the Battles of Cantigny, Aisne-Marne, St. Mihiel, and the Meuse-Argonne.

At the end of the war, Marshall, a colonel (temporary) served as aide-de-camp to General John J. Pershing, from 1919–1924. Reduced in grade to his permanent rank of lieutenant colonel (August 23, 1923), Marshall assumed command of the 15th Infantry Division in China (1924–1927). He returned to the United States and served as an instructor at the Army War College (1927), and then as an assistant commandant of the Infantry School at Fort Benning, Georgia (1927–1932). It was while at Fort Benning that Marshall came into contact with many of the officers who would lead the U.S. Army during World War II, including Majors Omar N. Bradley and Joseph Stilwell. Promoted to colonel (permanent) on September 1, 1933, Marshall then assumed command of the 8th Infantry Division, and then later as the senior instructor to the Illinois National Guard. On October 1, 1936, the Army advanced Marshall to the grade of brigadier general and assigned him to command the 5th Brigade, from 1936 to 1938. During his tour of duty with the 5th Brigade, a part of Marshall's duty was to manage the projects being worked on by the Civilian Conservation Corps, a joint Department of Labor and War Department project designed to employ unemployed youths in civilian conservation projects. Run by Army officers and noncommissioned officers alike, the program gave many young men an introduction to military life — this just prior to the outbreak of World War II, in September 1939.

In 1938, Brigadier General Marshall became chief of the War Plans Division of the Army General Staff (July–October 1938), and as deputy chief of staff, from October 1938 to July 1, 1939. On July 1, 1939, General (temporary) George C. Marshall became chief of staff of the Army, a position which he held until November 1, 1945. During World War II, General Marshall became the de facto chairman of the Joint Chiefs of Staff, and oversaw the prosecution of the U.S. war effort against both Nazi Germany and Imperial Japan. Over time, General Marshall became President Franklin D. Roosevelt's main spokesman at the major wartime conferences, including the Atlantic Charter Conference at Sea between Roosevelt and British Prime Minister Winston Churchill (August 1941); Casablanca Conference (January 1943); Cairo and Tehran Conferences (November 1943); Quebec Conference (September 1944); Yalta conference (February 1945); and Potsdam Conference (July 1945). General Marshall was advanced to the rank of five-star General of the Army (temporary) on December 16, 1944 (permanent on April 11, 1946). Through his immense skill as a manager and his professionalism as a soldier, General Marshall was the architect of the U.S. victory over the Axis.

General of the Army George C. Marshall retired from active military service on February 28, 1947. He continued to serve the nation as Secretary of State (1947–1949) and was the architect of the European Recovery Plan, later renamed the Marshall Plan. During the Korean War (June 1950–July 1953), Marshall served as secretary of war under President Harry S Truman (1950–1951) after the resignation of Louis Johnson in September 1950. The recipient of many honorary degrees, General Marshall was awarded the Nobel Peace Prize in 1953 for his Marshall Plan. He spent the remaining years of his life dedicated to the cause of peace and national security. General Marshall died on October 16, 1959, and was buried with full military honors, in Arlington National Cemetery.

Floyd Odlum, Husband of Jacqueline Cochran, 1892–1976

Floyd Odlum, husband of aviatrix Jacqueline Cochran, was born March 30, 1892, in Union City, Michigan. Odlum moved to New York City, where he worked as a corporate attorney. He later worked as a financier with such companies as Goldman Sachs and other financial institutions during the 1920s. During the decade of the Great Depression (1929–1938), he was one of a handful of men who made money on Wall Street, this despite the continued weakening of the U.S. economy. Besides his financial interests, he and several partners financed several Broadway Plays, including *The Pajama Game*. In May of 1932, he met aviatrix Jacqueline Cochran. With his help, she started a cosmetics company and a popular line of cosmetics. Divorcing his first wife Hortense McQuarrie, Odlum married Cochran on May 11, 1936, and encouraged her to pursue her interest in aviation. Cochran earned her pilot's license and entered several air races, including the Bendix Race in 1938, becoming the first female aviator to win that competition. The Odlum-Cochran home soon became a "Who's Who" in aviation and politics. Included among the friends and associates of the Odlums were Amelia Earhart and her husband-publisher George Putnam; then-Texas Congressman Lyndon Johnson and his wife, Lady Bird Johnson; aviator Chuck Yeager and his wife, Glennis; and after World War II, General (later President) Dwight D. Eisenhower. During World War II, with her husband's support and encouragement, Cochran founded the WASP (Women Airforce Service Pilots), which ferried newly built aircraft to airbases throughout the United States for the U.S. Army Air Forces.

After the war, Floyd Odlum continued his business pursuits, becoming president and chief executive officer of the Atlas Corporation in May of 1948. During the 1950s, Odlum and Cochran purchased a ranch in California, where Cochran continued to pursue her interests in aviation, setting several speed records in an F-86 Saber Jet, and later an F-104 Starfighter.

Diagnosed with rheumatoid arthritis during the 1940s, Odlum became chairman of the board of the Arthritis Foundation, which he co-founded in May 1948. During the 1940s and 1950s, he quietly backed Cochran in her aviation and political pursuits, though gradually retiring from public life during the same period. After several years of declining health, Floyd Odlum died on June 17, 1976, at the age of 84.

Alexander Nikolaivich Prokofieff de Seversky, Aeronautical Engineer, 1894–1974

Alexander Nikolaivich Prokofieff de Seversky was born in Triflis, Russia (now the Republic of Georgia) on June 7, 1894. Having grown up around aviation all his life, young Seversky had already flown before entering the Imperial Russian Naval Academy in 1910. After graduation in 1914, Seversky was commissioned a lieutenant and shortly thereafter entered flight training. In 1915, he entered the Imperial Russian Navy as an aviator. During one engagement he lost his right leg in a dogfight with a German aircraft; nevertheless, he was flying combat missions again, against German and Austro-Hungarian aviators, within the year. Flying 57 combat missions in all, Seversky became Russia's top naval air ace, shooting down six German airplanes.

Arriving in the United States as the Naval attaché to the New Russian Provisional Government under Alexander Kerensky, Seversky decided to remain in the United States after the February Revolution which brought Vladimir Illyich Lenin to power in 1917. Seversky immediately offered his services to the U.S. War Department, assisting in the design and construction of the British Se-5 fighter. At the end of the war, Seversky teamed up with Army aviator Brigadier General William E. "Billy" Mitchell in the setting up the bombing tests off Norfolk, Virginia, where Mitchell had hoped to demonstrate the effectiveness of air bombardment against ships of the Navy. Seversky likewise designed the system of in-flight refueling used by Army and Navy aircraft as well as filing some 364 patents, most of which dealt with aviation.

Leaving government service for a while, Seversky went to work for the Sperry Gyroscope Company, where he designed and built the gyro-stabilized bombsight that was hailed as a milestone in the development of aerial bombardment. Seversky, now a major in the U.S. Army Air Corps Reserve, founded the Seversky Aircraft Corporation, where he and his fellow engineers worked on several aircraft designs, the most famous being the P-47 "Thunderbolt" or "Jug" (as it was affectionately called by Army pilots who flew her during World War II). When the stock market crashed in October 1929, Seversky's corporation went bankrupt. Bouncing back in 1931, Seversky and a group of investors founded the Seversky Aircraft Corporation in New York City, in February 1931. One of the first planes built by Seversky's corporation was an all-metal skin amphibious monoplane, the SEV-3. Joining Seversky's firm was Russian émigré Alexander Kartveli, who later designed the P-47 Thunderbolt, a rugged close air support fighter-bomber aircraft used by Army Air Force pilots in the Mediterranean, European, and Pacific theaters. Seversky's company likewise designed the P-35 and later the P-36. The Seversky Aircraft Corporation once again fell on hard times. Instead of filing for bankruptcy, the managing director, W. Wallace Kellett, managed to remove Seversky from his own company. In October 1939, Kellett renamed the company The Republic Aviation Corporation.

Forced out of his business, Seversky then turned to writing and consulting. In 1942, he authored a book, alerting the nation to the need for better air power, entitled *Victory Through Air Power*, and it immediately became a bestseller. In the book, Seversky asserted that air power was the key to victory, and that the airplane had made traditional forms of land and sea warfare obsolete. Seversky emphasized that the current war had proved that warfare was revolutionary and, as such, "revolutionary responses" were required. He also put forth the argument that American fighters and bombers were obsolete as compared to those of Germany, Japan, and Italy, a claim repeatedly denied by the leaders of the Army Air Force and American press.

After the war, Seversky was honored by President Harry S Truman, who awarded him the Medal of Merit. After the founding of the U.S. Air Force in 1947, Alexander Seversky served as a consultant with the new organization. He also served as a special consultant to the Chiefs of Staff of the U.S. Air Force on air power for the next two decades. For his work with the U.S. Air Force, Severesky was awarded the Exceptional Service Medal in 1969. Always interested in technology, he formed the Seversky Electroatom Corporation, a company that focused on protecting the United States from a nuclear attack by the Soviet Union and on extracting radioactive particles from the air.

Alexander Seversky died on August 24, 1974. He will be forever remembered for the work he did on the gyro-stabilizer and the in-flight refueling technology he developed in the early 1920s, and of his advocacy for strategic bombing.

Lieutenant General Barton Kyle Yount, USAF, 1884–1949

Lieutenant General Barton Kyle Yount, USAF, was born in Troy, Ohio, on January 18, 1884. After attending high school there he entered Ohio State University (1902–1903) prior to his acceptance into the United States Military Academy at West Point, New York, in 1903. Commissioned in 1907 in the Infantry, Yount reported to the 27th Infantry Division during the Army's pacification campaign there (1906–1909). He served at Fort Sheridan, Illinois, from April 1909 to February 1913. He then served with the 4th Brigade, 2nd Division, stationed at Texas City, Texas, until July 1914. Leaving Texas, Yount reported to Tientsin, China, for duty with the 15th Infantry Regiment, until August 1917. Transferred to the Signal Corps for aviation duty, then Major (temporary) Yount reported to the Aviation Section, where he remained until September 1917, at which time he reported to Kelly Field, Texas. He was soon reassigned to Austin, Texas, where he served as an instructor in military aeronautics. He later served on a board of Army officers, overseeing the reorganization of the Army Air Service. In July 1919, he successfully completed aviation training duty and was assigned as commanding officer of March Field, California, from July 1919 until the summer of 1921, whereupon he reported to Washington, D.C., where he served for three years in the Office of the Chief of the Air Corps. In August 1924, Yount enrolled in the Air Corps Engineering School at McCook Field, Ohio.

Upon graduation from the Air Corps Engineering School, McCook Field, he sailed for France in the summer of 1925, where he served for four years as the American Military Attaché for Aviation in Paris. During this period he also served as a delegate at several international aviation conferences; as a Technical Expert at the Disarmament Conference at Geneva, Switzerland; and attended the Extraordinary Session of the International Commission for Air Navigation in June 1929.

Returning to the United States in August 1929, he attended the Air Corps Tactical School, Langley Field, Virginia, for a year. Upon graduation from the Air Corps Tactical School, he commanded the Rockwell Air Depot, in Coronado, California, until July 1932, and Bolling Field, D.C., until July 1934. He was a member of a board of officers, convened in the Office of the Chief of Air Corps, Washington, D.C., July–August 1934, for the purpose of determining the location of the Air Corps Technical School. He then attended the Army Industrial College, D.C., and graduated from there in August 1936. He enrolled in the Army War College, also located in Washington, D.C., graduating in June 1936.

The following August he sailed for Hawaii, where he was assigned to duty as commanding officer, 18th Composite wing, with headquarters located at Fort Shafter, Territory of Hawaii. He returned to the United States in July 1938 and assumed command of the Air Corps Training Center at Randolph Field, Texas. In February 1939, he was transferred to Washington, D.C., for duty once again with the Office of the Chief of the Air Corps as Chief of the Army Air Corps' Training and Operations Division, and was put in charge of all training activities of the Army Air Corps.

In October 1943, Yount was assigned to the Panama Canal Department Air Force, and, in November 1940, was placed in command of the

Southeast Air District, with its headquarters in Tampa, Florida. He was then assigned to command the West Coast Air Corps Training Center, Moffett Field, California, in July 1941. With the U.S. entrance into World War II, Major General Yount reported to Washington, D.C., in January 1942, for duty in the Office of the Chief of the Air Corps. In March 1942, he became commanding general of the Army Air Forces flying Training Command, Fort Worth, Texas. General Yount recommended the training of the Women Airforce Service Pilots (WASP) under the auspices of the Army Flying Training Command, and in fact, accompanied WASP founder Jacqueline Cochran on an inspection tour in December 1943 of the first class of Women pilots. Impressed with the work that Cochran was doing, General Yount later reported to Army Air Force officials that the women pilots were "outstanding." Promoted to lieutenant general, Yount remained in his capacity of Army Air Force training until the end of the war. Retiring on July 1, 1946, Yount spent the remaining years in Phoenix, Arizona, where he served as president of the American Institute of Foreign Trade, until his death on July 11, 1949. After a brief funeral service in the chapel at Fort Myer, Virginia, Lieutenant General Yount was laid to rest, with full military honors, in Arlington National Cemetery. His wife, Mildred, is buried beside him, as is his son, Barton K. Yount, Jr., who as an Air Force colonel, was killed in action while on a combat mission in the Republic of Vietnam, in February 1969.

General Yount untiringly contributed to the training of both officers and enlisted men for the Army Air Forces. It was to his credit that the Army Air Forces became one of the cornerstones of victory in the air war over Germany and Japan during World War II.

Appendix Two

Major Aircraft Flown by Jacqueline Cochran

Pre–World War II: Seversky P-35
1935–1940 "Gee Bee" Sportster R-1
 Boeing B-17 Bomber

World War II: Curtiss P-40 "Warhawk"
1941–1945 Republic P-47 "Thunderbolt"
 P-38 Lockheed "Lightning"
 B-25 North American Bomber
 B-26 Martin "Marauder"
 P-51 North American "Mustang"
 B-24 Consolidated "Liberator"
 Douglas C-47 Transport
 P-39 "Air Cobra"
 AT-6 Trainer

Post–World War II: P-51 "Super Mustang"
1946–1964 F-86 North American Sabrejet Jet Fighter
 F-104 Lockheed "Starfighter"

Chapter Notes

Introduction

1. *Fly Girls: Female Pilots in World War II*, Documentary Video, KET-TV, originally aired May 24, 1999.
2. "Jacqueline Cochran: Breaking Aviation Records is Only One of her Many Activities," *U.S. Air Services*, Loose Article, January 1938, Cochran, Jacqueline, Documents, Vertical Files, CC-344500-01, National Air and Space Museum, Washington D.C.
3. Jackie Cochran and Maryann Bucknum Brinley, *Jackie Cochran: The Autobiography of the Greatest Woman Pilot in Aviation History* (Toronto and New York: Rufus Publications, 1987), p. 121.
4. General Chuck Yeager and Leo Janos, *Chuck Yeager: An Autobiography* (New York: Bantam, 1985), p. 268.
5. *Fly Girls*, Documentary Video, 1999.
6. Susan Ware, *Still Missing: Amelia Earhart and the Search for Modern Feminism* (New York and London: W.W. Norton, 1993), p. 129.
7. Yeager and Janos, *Chuck Yeager*, p. 268.
8. See for example, Allen F. Davis, *American Heroine: The Legend of Jane Adams* (New York: Oxford University Press, 1973), Lois Banner, *Elizabeth Cady Stanton: A Radical for Woman's Rights* (Boston: Little, Brown, 1980); or, Blanche Weisen Cook, *Eleanor Roosevelt* (New York: Viking, 1992).
9. Ware, *Still Missing: Amelia Earhart and the Search for Modern Feminism*, p. 176.
10. Mary S. Lovell, *The Sound of Wings: The Life of Amelia Earhart* (New York: St. Martin's Press, 1989); and Ware, *Still Missing: Amelia Earhart and the Search for Modern Feminism*.
11. Joseph J. Corn, *The Winged Gospel: America's Romance with Aviation 1900-1950* (New York and Oxford: Oxford University Press, 1983). For a basic introduction to American women in aviation, see: Kathleen Brooks-Pazmany, *United States Women in Aviation 1919-1929* (Washington and London: Smithsonian Institution Press, 1991); Claudia M. Oakes, *United States Women in Aviation 1930-1939* (Washington and London: Smithsonian Institution Press, 1991); and Deborah G. Douglas, *United States Women in Aviation 1940-1985* (Washington and London: Smithsonian Institution Press, 1991).
12. D'Ann Campbell, "Women in Uniform: The World War II Experiment," *Military Affairs* 51 (July 1987), p. 137.
13. Mattie E. Treadwell, *United States Army in World War II: Special Studies, The Women's Army Corps* (Washington D.C.: Department of the Army, 1954); D'Ann Campbell, *Women at War with America: Private Lives in a Patriotic Era* (Cambridge, MA. and London, England: Harvard University Press, 1984); and Jeanne Holm, *Women in the Military: An Unfinished Revolution* (Novato, CA: Presidio Press, 1982). For women in industry, see: Ruth Milkman, *Gender at Work: The Dynamics of Job Segregation by Sex during World War II* (Urbana and Chicago: University of Illinois Press, 1987). For an overall look at women in World War II, see: Karen Anderson, *Wartime Women: Sex Roles, Family Relation, and the Status of Women During World War II* (Westport, CT: Greenwood Press, 1981); and Susan M. Hartmann, *The Home Front and Beyond: American Women in the 1940s* (Boston: Twayne, 1982). For a recent review of books on women in the mil-

itary, see: D'Ann Campbell, "Women in the Military," *Choice* 31 (September, 1993), pp. 63–70.

14. Byrd Howell Granger, *On Final Approach: The Women Airforce Service Pilots of W.W. II* (Scottsdale, AZ: Falconer, 1991); Adela Riek Scharr, *The WASPs*, Vol. II: *Sisters in the Sky* (St. Louis, MO: Patrice Press, 1988); Marianne Verges, *On Silver Wings: The Women Airforce Service Pilots of World War II 1942–1944* (New York: Ballantine Books, 1991); and Jean Hascall Cole, *Women Pilots of World War II* (Salt Lake City: University Utah Press, 1992).

15. Molly Merryman, *Clipped Wings: The Rise and Fall of the Women Airforce Service Pilots (WASPs) of World War II* (New York and London: New York University Press, 1998).

16. See Jacqueline Cochran and Maryann Bucknum Brinley, *Jackie Cochran: The Autobiography of the Greatest Woman Pilot in Aviation History* (Toronto and New York: Rufus, 1987).

17. See Jacqueline Cochran, *The Stars at Noon* (Boston and Toronto: Little, Brown, 1954).

Chapter 1

1. Cochran, *The Stars at Noon*, p. 3.
2. Ibid., p. 4.
3. *Florida Gazetteer and Business Directory 1911–1912* (Jacksonville, Florida: R.L. Polk) p. 393, State Library of Florida, Tallahassee, Florida.
4. Cochran, *The Stars at Noon*, pp. 5, 6; and Cochran and Brinley, *Jackie Cochran: The Autobiography*, pp. 7, 8.
5. Cochran, *The Stars at Noon*, p. 3.
6. Ibid., pp. 20, 21; and Cochran and Brinley, *Jackie Cochran: The Autobiography*, p. 348.
7. Cochran, *The Stars at Noon*, pp. 13, 14, 15.
8. Ibid., p. 20; and Brinley and Cochran, *Jackie Cochran: The Autobiography*, p. 30.
9. Cochran, *The Stars at Noon*, pp. 17, 18.
10. Ibid., p. 17.
11. Ibid., pp. 17, 18.
12. Ibid., pp. 18, 19.
13. Ibid., p. 11.
14. Ibid., pp. 21, 22.
15. Ibid., p. 22.
16. For more information on the cotton industry and child labor, see Broadus Mitchell, *The Rise of the Cotton Mills in the South* (New York: Da Capo Press, 1968); and Jacqueline Dowd Hall, et al., *Like a Family: The Making of a Southern Cotton Mill World* (New York and London: W.W. Norton, 1987).
17. Cochran, *The Stars at Noon*, pp. 23, 24.
18. Ibid., p. 24.
19. Cochran and Brinley, *Jackie Cochran: The Autobiography*, p. 34; and Cochran, *The Stars at Noon*, p. 24.
20. Cochran and Brinly, *Jackie Cochran: The Autobiography*, p. 35.
21. Cochran, *The Stars at Noon*, p. 22.
22. Ibid., p. 26.
23. Ibid., pp. 26, 27.
24. Hall, et al., *Like a Family: The Making of a Southern Cotton Mill World*, p. 58.
25. Cochran, *The Stars at Noon*, pp. 28, 29.
26. Ibid., p. 29.
27. Ibid., pp. 27, 30, 31.
28. Ibid., pp. 30, 31.
29. Cochran and Brinley, *Jackie Cochran: The Autobiography*, p. 43.
30. Cochran, *The Stars at Noon*, p. 31.
31. The Catholic hospital Cochran attended was most probably St. Margaret's Hospital, which opened in 1902, and later became part of Humana Hospital. The nursing program was likely directed by Catholic nuns from the Order of the Daughters of Charity. This information came from a telephone interview with Ed Stugardi, Administrator, St. Providence Hospital (Montgomery, AL), and Sister Margaret Flynn, Order of the Daughters of Charity (Evansville, IN), December 21, 1995.
32. Cochran, *The Stars at Noon*, p. 31.
33. Cochran and Brinly, *Jackie Cochran: The Autobiography*, p. 44.
34. Ibid., pp. 44, 45.
35. Cochran, *The Stars at Noon*, p. 32.
36. In her 1954 autobiography, Cochran fails to mention how long she worked as a nurse. Her time in Bonifay was less than a year, and probably only a few months. When Cochran went into nursing, airlines preferred to hire nurses as stewardess, which gave passengers a feeling of comfort and safety. In fact, a nursing degree was often a requirement for employment as an airline hostess.

This idea was first suggested due to airsickness among passengers. Cochran, *The Stars at Noon*, pp. 32, 33; and Roger E. Bilstien, *Flight in America: From the Wright Brothers to the Astronauts* (Baltimore and London: Johns Hopkins University Press, 1984), pp. 101, 102.

37. Cochran, *The Stars at Noon*, p. 32.
38. Ibid., p. 33.
39. Enid Shomer, *Stars at Noon: Poems from the Life of Jacqueline Cochran* (Fayetteville: University of Arkansas Press, 2001), p. 3.
40. For a discussion on social feminists' activities after the 19th amendment, see: J. Stanley Lemons, *The Women Citizen: Social Feminism in the 1920s* (Urbana, Chicago and London: University of Illinois Press, 1973).
41. Dorothy Brown, *Setting a Course: American Women in the 1920s* (Boston: Twayne, 1987), p. 29.
42. Nancy Woloch, *Women and the American Experience: A Concise History* (New York: McGraw-Hill, 1996), p. 241.
43. Ibid., p. 248.
44. Glenda Riley, *Inventing the American Woman: A Perspective on Women's History 1865 to the Present* (Arlington Heights, IL: Harlan Davidson, 1986), p. 67.
45. Brown, *Setting a Course*, p. 33.
46. Ibid.
47. Oral History Interview of Miss Jacqueline Cochran by Captain Robert S. Bartanowicz and Major John "Fred" Shiner, 11–12 March 1976, Typed Transcript, [8], K239.0512-940, in USAF Collection, AFHRA.
48. For a discussion on young people during the 1920s, see Paula S. Frass, *The Damned and the Beautiful: American Youth in the 1920s* (Oxford: Oxford University Press, 1977). The book provides a contrast to Cochran's life since it discusses young people in college.
49. Cochran, *The Stars at Noon*, p. 33.
50. Ibid., p. 34.
51. Cochran and Brinley, *Jackie Cochran: The Autobiography*, p. 51.
52. Cochran, *The Stars at Noon*, p. 34; and Cochran and Brinly, *Jackie Cochran: The Autobiography*, p. 51.
53. Woloch, *Women and the American Experience*, p. 249.
54. Riley, *Inventing the American Woman*, p. 86.
55. Cochran, *The Stars at Noon*, pp. 35, 36, 37.
56. Ibid., pp. 37, 38.
57. Cochran and Brinley, *Jackie Cochran: The Autobiography*, p. 118.

Chapter 2

1. Brown, *Setting a Course*, p. 42.
2. Bilstien, *Flight in America*, p. 29.
3. Ibid., p. 36.
4. Ibid., p. 41.
5. Ibid.
6. Ibid., p. 83
7. Corn, *The Winged Gospel*, p. 12.
8. Lloyd Morris and Kendall Smith, *Ceiling Unlimited: The Story of American Aviation from Kitty Hawk to Supersonics* (New York: Macmillan, 1953), pp. 227, 228.
9. Ibid., p. 227.
10. Bilstein, *Flight in America*, p. 59.
11. Ibid., pp. 50, 51.
12. Elihu Rose, "The Court-Martial of Billy Mitchell," *The Quarterly Journal of Military History—Special Issue: Air Power* (Spring 1996), p. 17.
13. Morris and Smith, *Ceiling Unlimited: The Story of American Aviation from Kitty Hawk to Supersonics*, p. 230.
14. Corn, *The Winged Gospel*, p. 25.
15. Bilstein, *Flight in American: From the Wrights to the Astronauts*, p. 76.
16. Hendrick de Leeuw, *Conquest of the Air: The History and Future of Aviation* (New York: Vantage Press, 1960), pp. 88, 89.
17. Bilstein, *Flight in America*, p. 84.
18. de Leeuw, *Conquest of the Air*, p. 140.
19. "Propeller Annie," *Aviation Quarterly* 5 (1979), p. 325.
20. Mary Cadogan, *Women with Wings: Female Flyers in Fact and Faction* (Chicago: Academy Chicago, 1993), pp. 39, 33.
21. Donald S. Lopez, *Aviation: A Smithsonian Guide* (New York: Macmillan, 1995), p. 64.
22. Charles E. Planck, *Women with Wings* (New York and London: Harper and Brothers, 1942), pp. 21, 22.
23. Cadogan, *Women with Wings: Female Flyers*, p. 16.
24. Lopez, *Aviation: A Smithsonian Guide*, p. 64.
25. Corn, *The Winged Gospel*, p. 73.
26. Untitled Biographical Information on Ruth Rowland Nichols, Ruth Rowland Nichols, Vertical Files, CN-06000-01, Na-

Notes — Chapter 2

tional Air and Space Museum, Washington, D.C.; and "An American Aviatrix Abroad: Miss Ruth Nicholas Flies and Studies Aviation on Trip Around the World," *Aero Digest*, September 1925, Ruth Rowland Nichols, Vertical Files, CN-06000-01, National Air and Space Museum, Washington, D.C.

27. "Women's Progress in Air is Shown By Naming of 3 as 'Marking Pilots,'" Newspaper Clipping, October 5, 1935, Louise McPhetridge Thaden (1989-0132), Vertical Files, CT-141000-03, National Air and Space Museum, Washington D.C.; and "Air and Space Museum biog.," Louise McPhetridge Thaden (1989-0132), Vertical Files, CT-141000-01, National Air and Space Museum, Washington, D.C.

28. "Laura Ingalls' Desire to Fly Amused Pilots," *Washington Herald* (September 13, 1935); and "Miss Laura Ingalls: First American Woman to Fly the Andes," *US Air Services: Feature Aeronautical Magazine Commercial and Military* (June 1934), no page numbers, in Laura A. Ingalls, Vertical Files, CI-020020-01, National Air and Space Museum, Washington, D.C.; and "Propeller Annie," *Aviation* 5 (Fourth Quarter, 1979), in Helen Richey, Vertical Files, CR-3511000-01, National Air and Space Museum, Washington, D.C.

29. Corn, *The Winged Gospel*, p. 73.
30. Ware, *Still Missing*, p. 61.
31. Corn, *The Winged Gospel*, p. 76.
32. Ibid., p. 75.
33. Ibid., p. 76.
34. Ibid.
35. Ware, *Still Missing*, p. 62.
36. Vera L. Connolly, "Daughters of the Sky," *The Delineator* 115 (August, 1929), p. 9.
37. Ibid., pp. 82, 83.
38. Ibid., p. 82.
39. Ware, *Still Missing*, p. 129.
40. Ibid., p. 66.
41. Amelia Earhart, *The Fun of It: Random Records of My Own Flying and Women in Aviation* (Chicago: Academy Chicago, 1932), p. 106.
42. Cochran and Brinley, *Jackie Cochran: The Autobiography*, p. 84.
43. Ibid., p. 83.
44. Sherryl Connelly, "Just Plain Great: Jackie Cochran had the stuff Legends are made of," Newspaper Clipping, July 23, 1987, Cochran, Jacqueline Documents, CC-344500-01, National Air and Space Museum, Washington, D.C.

45. Cochran and Brinley, *Jackie Cochran: The Autobiography*, p. 94.
46. Lois Banner, *Women in Modern America: A Brief History* (San Diego: Harcourt Brace Jovanovich, 1984), p. 183.
47. Cochran and Brinley, *Jackie Cochran: An Autobiography*, p. 57.
48. It is difficult to pinpoint exactly when Jacqueline Cochran first considered becoming a pilot. In a 1938 interview she told WMCA radio in New York that she first thought about flying as a child in Pensacola while watching Navy airplanes. Yet she fails to mention this incident in either her 1954 autobiography or the Cochran/Brinley 1987 autobiography. In the 1954 biography Cochran claims she first considered the idea of flying that evening when Odlum made his suggestion. Cochran, *The Stars At Noon*, p. 40. And Jacqueline Cochran to Frankie Basch, "Success Stories," WMCA Radio Broadcast Transcript, April 12, 1938, Talks and Speeches 1938 (2), General File #5, Jacqueline Cochran Collection, Dwight D. Eisenhower Presidential Library, Abilene, KS.
49. Cochran and Brinley, *Jackie Cochran: The Autobiography*, p. 12.
50. Ibid., pp. 58, 59.
51. Cochran had bet Floyd Odlum that she could earn a pilot's license in three weeks' time. Ibid, p. 71.
52. Ibid.
53. Cochran, *The Stars at Noon*, pp. 42, 43; Cochran and Brinley, *Jackie Cochran: The Autobiography*, pp. 71, 72; "Girl Wins Bet by Earning Pilot's License During Her Vacation of Three Weeks," *New York World Telegram* (August 18, 1932), Scrapbook 1932–1937, Scrapbook Series Box #19, Jacqueline Cochran Collection, Dwight D. Eisenhower Presidential Library, Abilene, KS.
54. Cochran and Brinley, *Jackie Cochran: The Autobiography*, p, 72.
55. "Girl Wins Bet by Earning Pilot's License During Her Vacation of Three Weeks," *New York World* (August 18, 1932).
56. Cochran, *The Stars at Noon*, p. 43.
57. Ibid.
58. Ibid., pp. 43, 44.
59. Cochran and Brinley, *Jackie Cochran: The Autobiography*, p. 80.

60. Cochran, *The Stars at Noon*, p. 45. And Cochran and Brinley, *Jackie Cochran: The Autobiography*, p. 84.
61. Cochran, *The Stars at Noon*, p. 50.
62. Cochran and Brinley, *Jackie Cochran: The Autobiography*, pp. 97, 98, 99, 100; and Cochran, *The Stars at Noon*, p. 51
63. Jackie Cochran, "My Flying," Stories Written By Me—Jackie (1), Article Series, Box #1, p. 12, Jacqueline Cochran Collection, Dwight D. Eisenhower Presidential Library, Abilene, KS.
64. Cochran and Brinley, *Jackie Cochran: The Autobiography*, p. 93.

Chapter 3

1. Elizabeth A. Valentine, "No. 1 Woman Flier: Jacqueline Cochran, Who Flew a Bomber to Britain, Advises American Women to Follow a British Example," *New York Times*, July 13, 1941, p. 11.
2. Bartanowicz and Shiner, Interview of Miss Jacqueline Cochran, p. 4.
3. "Aviatrix Wrecks Plane," *New York Times*, May 14, 1934, p. 3.
4. Cochran, *The Stars at Noon*, p. 50.
5. "Miss Cochran Set Woman's Air Mark," *New York Times*, September 21, 1937, p. 29; and "Records Shattered at Miami Meet," *The Bee Hive*, December, 1937 (Clipping, no page number), in Jacqueline Cochran, Vertical Files, CC-344500-02, National Air and Space Museum, Washington, D.C.
6. Rochelle Chadakoff, ed. *Eleanor Roosevelt's My Day: Her Acclaimed Columns 1936–1945* (New York: Pharos Books, 1989), p. 138.
7. "Miss Cochran 33,000 Feet Up," *New York Times*, March 25, 1939, p. 17.
8. "Jacqueline Cochran: Achieving Heights in Flying and Cosmetics," Fact Sheet, Cochran, Jacqueline Documents, Vertical Files, CC-344500-01, National Air and Space Museum, Washington, D.C.
9. Richard P. Hallion, *Test Pilots: The Frontiersmen of Flight* (Washington D.C. and London: Smithsonian Institute Press, 1981, 1988), p. 295.
10. Cochran, *The Stars at Noon*, p. 50.
11. Ibid., p. 62.
12. Ibid.
13. Jacqueline Cochran to George M. Lewis, August 1, 1940, General Files #8, Collier Trophy 1940 (3), Jacqueline Cochran Collection, Dwight D. Eisenhower Presidential library, Abilene, KS.
14. Cochran and Brinley, *Jackie Cochran: The Autobiography*, p. 131.
15. For more information on aviation medicine research see Harry G. Armstrong, *Principles and Practice of Aviation Medicine* (Baltimore: Williams and Wilkins, 1943).
16. Cochran and Brinley, *Jackie Cochran: The Autobiography*, pg. 131.
17. Ibid.
18. Cochran and Brinley, *Jackie Cochran: The Autobiography*, p. 146.
19. Alexander P. De Seversky, *Victory Through Air Power* (New York: Simon & Schuster, 1942), pp. 231, 232.
20. Ibid., p. 232.
21. Cochran and Brinley, *Jackie Cochran: The Autobiography*, p. 146.
22. Ibid., p. 148.
23. Ibid., p. 146.
24. Ibid., 146, 147.
25. Ibid., p. 148.
26. Robert E. Bail, "Record Set By Woman: Miss Cochran Made World Speed Mark in Plane Strange to Her," *New York Times*, September 26, 1937, p. 4.
27. Cochran and Brinley, *Jackie Cochran: The Autobiography*, p. 150.
28. Morris and Smith, *Ceiling Unlimited: The Story of American Aviation from Kitty Hawk to Supersonics*, pp. 93, 98.
29. "Aviation: Earth's Distance Shrivel as Melbourne Race Flyers Cut Air Travel from Britain to Australia to Two Days," *News-Week*, 4 (October 27, 1934), p. 5.
30. "The MacRobertson International Air Races: England—Australia," Conditions. File: London Race General (9), Box: Annual Series #2, Jacqueline Cochran Collection, Dwight D. Eisenhower Presidential Library, Abilene, KS; and Christopher Chant. *Aviation: An illustrated History* (London: Orbis, 1978), p. 160.
31. "The MacRobertson International Air Races: England—Australia," Conditions. File: London Race General (9), Box: Annual Series #2, Jacqueline Cochran Collection, Dwight D. Eisenhower Presidential Library, Abilene, KS. And Jacqueline Cochran, "London-Melbourne Air Race," London Flight General (10), Annual Series Box #2, Jacque-

Notes — Chapter 3

line Cochran Collection, Dwight D. Eisenhower Presidential Library, Abilene, KS.

32. Ibid., p. 52.

33. Charles A. Mendenhall and Tom Murphy, *The Gee Bee Racers: A Legacy of Speed* (North Branch, MN: Specialty Press), p. 7.

34. Cochran, *The Stars at Noon*, p. 55.

35. Cochran's airplane was somewhat of an oddity at the MacRobertson Race. The British referred to the Gee Bee as the "HeeBeeGeeBee," or simply "ugly and dumpy." The plane's "Lucky Strike Green" paint job did nothing to enhance its appearance. Mendenhall and Murphy, *The Gee Bee Racers*, p. 103; and Cochran and Brinley, *Jackie Cochran: The Autobiography*, p. 108.

36. According to Jacqueline Cochran, she bought the Gee Bee after Royal Leonard crashed the Gamma. Cochran, *The Stars at Noon*, p. 53. Yet, according to another source, Cochran bought the "Gee Bee" as a backup in case the Gamma's supercharger failed. Mendenhall and Murphy, *The Gee Bee Racers: A Legacy of Speed*, p. 97. It is more likely that Cochran bought the Gee Bee after the Gamma failed. After all, she spent a considerable amount of money to rebuild the Gamma and felt overconfident about winning. It is unlikely she would have provided a backup plane, and more probable that she told the story about needing a backup to avoid telling the Granville Brothers that her supercharged Gamma had crashed.

37. Cochran did not provide an exact description of the plane's condition. Right before their test run, the plane had not been tuned or tested and was missing the rear seat. It was also unfinished mechanically. Cochran, *The Stars at Noon*, p. 53.

38. Cochran, *The Stars at Noon*, p. 53.

39. Cochran, *The Stars at Noon*, pp. 53, 54; and "Scott and Black Win, Flying to Melbourne in 71 Hours; Dutch Still Second; 2 Killed," *New York Times*, October 28, 1934, p. 1.

40. Cochran, *The Stars at Noon*, p. 55.

41. Jacqueline Cochran, "The Bendix Air Race," Miscellaneous Aviation Speeches (2), Speech Series, Box #1, 2 Jacqueline Cochran Collection, Dwight D. Eisenhower Presidential Library, Abilene, KS.

42. Cochran, "The Bendix Air Race," p. 3.

43. Ibid., p. 2.

44. Ibid.

45. Glines, *Roscoe Turner: Aviation's Master Showman*, p. 224.

46. Cochran, *The Stars at Noon*, pp. 62, 63.

47. Ibid., p. 64.

48. Cochran and Brinley, *Jackie Cochran: The Autobiography*, p. 97.

49. "Frank Fuller, The Man to Beat in the Bendix." Newspaper Clipping, September 2, 1938, Scrapbook 1938 (3), Scrapbook Series Box #9, Jacqueline Cochran Collection, Dwight D. Eisenhower Presidential Library, Abilene, KS.

50. Cochran, "The Bendix Race," p. 9.

51. Ibid., p. 10.

52. Ibid., p. 11.

53. Ibid., p. 10.

54. Cochran and Brinley, *Jackie Cochran: The Autobiography*, p. 161.

55. "Corrigan Day" is a reference to Douglas "Wrong-way" Corrigan, who in 1938 accidentally flew to Ireland on his way to California. de Leeuw, *Conquest of Air*, p. 113; and Cochran, "The Bendix Race," p. 12.

56. "Jacqueline Cochran Takes First Money in Bendix Air Race Over 10 Male Pilots," *Santa Rosa California Republican*, September 3, 1938, Scrapbook 1938 (3), Scrapbook Series Box #19, Jacqueline Cochran Collection, Dwight D. Eisenhower Presidential Library, Abilene, KS. And "Miss Cochran First at Bendix," *New York Times*, September 4, 1938, p. 1.

57. Cochran, "The Bendix Race," p. 12.

58. "Woman Victor in Bendix Air Race; "Beat Eight Men," *Rock Island Ill Argus*, September 3, 1938, Scrapbook 1938 (3), Scrapbook Series Box #19, Jacqueline Cochran Collection, Dwight D. Eisenhower Presidential Library, Abilene, KS.

59. Ibid.

60. "Jacqueline Cochran," Unidentified Information Sheet, Talks and Speeches 1938 (12), General Files #5, Jacqueline Cochran Collection, Dwight D. Eisenhower Presidential Library, Abilene, KS.

61. "Interview — Miss Graham "America Weekly" November 12, 1948, Rough Draft of an article, Articles #1 Jacqueline Cochran — Articles 1948, Jacqueline Cochran Collection, Dwight D. Eisenhower Presidential Library, Abilene, KS.

62. "Aviation Award Won by Woman Flier," *New York Times*, February 19, 1938, p. 17; "Jacqueline Cochran Gets Trophy," *New York Times*, April 5, 1938, 19; and "Miss Cochran Gets Trophy Again," *New York Times*, June 16, 1939, p. 14.
63. Corn, *The Winged Gospel*, p. 72.
64. Dwiggins, *They Flew the Bendix*, p. 79.
65. Ibid.
66. Ibid., p. 88.
67. Louise Thaden, *High, Wide and Frightened* (New York: Stackpole and Sons, 1938), p. 258.
68. Charles E. Planck, *Women with Wings* (New York and London: Harper and Brothers, 1942), p. 90.
69. Susan Ware, *Still Missing*, p. 82.
70. Amelia Earhart, *The Fun of It*, p. 154.
71. Lopez, *Aviation: A Smithsonian Guide*, p. 69.
72. Ibid.
73. Ibid.
74. Earhart, *The Fun of It*, p. 152.
75. "Happiness, Horror Mingle, Women Derbyists Fly On," August 25, 1929, Newspaper Clipping, Louise McPhertridge Thaden (1989-0132), Vertical Files, CT-141000-02, National Air and Space Museum, Washington, D.C.
76. Planck, *Women with Wings*, p. 81.
77. Corn, *The Winged Gospel*, p. 84.
78. "Airplane Races Too Hazardous an Adventure for Women Pilots," *New York American* August 27, 1929, In Louise Thaden Vertical Files, CT-141000-02, National Air and Space Museum, Washington, D.C.
79. Earhart, *The Fun of It*, p. 152.
80. "Jacqueline Cochran: A Renaissance Woman for the 20th Century," *The Retired Officer* (September 1971), 30, M1 Cochran, Jacqueline, USAF Museum, Wright-Patterson AFB, Dayton, Ohio.
81. Cochran, Jacqueline, "I Reached the Stars the Hard Way." *Life*, August 16, 1954, p. 107.
82. Jacqueline Cochran to Miss Graham (no other name given), Interview for *American Weekly*, November 12, 1948, 11, Articles Series, Box #1, Jacqueline Cochran Collection, Dwight D. Eisenhower Presidential Library, Abilene, KS.
83. Bartanowicz and Shiner, Oral History Interview of Miss Jacqueline Cochran, p. 4.

84. Cochran, *The Stars at Noon*, p. 62.
85. Ware, *Still Missing*, p. 80.
86. Untitled Interview between Jacqueline Cochran and Mr. Uttal, National Air Races 1938 Miscellaneous, General Files #5, Jacqueline Cochran Collection, Dwight D. Eisenhower Presidential Library, Abilene, KS.
87. Ibid.
88. Ibid.
89. Ibid.
90. "Jacqueline Cochran Takes First Money in Bendix Air Race Over 10 Male Pilots," *Santa Rosa Republican*, Sept. 3, 1938, Scrapbook 1938 (3), Scrapbook Series Box #19, and "Jacqueline Cochran Win Bendix Derby," *Dallas Times-Herald*, Sept. 4, 1938, Scrapbook 1938 (3), Scrapbook Series Box #19, Jacqueline Cochran Collection, Dwight D. Eisenhower Presidential Library, Abilene, KS.
91. "Jacqueline Cochran Wins Bendix Derby," *Dallas Times-Herald*, September 4, 1938, no page number, in Cochran, Scrapbook Series, Box #19, Jacqueline Cochran Collection, Dwight D. Eisenhower Presidential Library, Abilene, KS.
92. Devon Francis, "Jackie Cochran Gets What She Wants and How," *Ashville Times*, Sept. 25, 1938, Scrapbook 1938 (3), Scrapbook Series, Box #19, Jacqueline Cochran Collection, Dwight D. Eisenhower Presidential Library, Abilene, KS.
93. Dwiggins, *They Flew the Bendix*, p. 105.
94. Ibid.
95. "Universal, Monday 22, (October) 1934," Press Release, London Flight General (8), Annual Series Box #2, Jacqueline Cochran Collection, Dwight D. Eisenhower Presidential Library, Abilene, KS.
96. Corn, *The Winged Gospel*, p. 84.
97. Ibid, p. 65.
98. Corn, *The Winged Gospel*, p. 83.
99. Cochran, *The Stars at Noon*, p. 65.

Chapter 4

1. Cochran and Brinley, *Jackie Cochran: The Autobiography*, p. 61.
2. "Mrs. Odlum Gets Nevada Divorce," *New York Times*, October 8, 1935, p.19.
3. Cochran and Brinley, *Jackie Cochran: The Autobiography*, p. 69.
4. *Fly Girls*, Documentary Video, 1999.

Notes — Chapter 4

5. Ibid.
6. Cochran, *The Stars at Noon*, Preface.
7. "Why do You Adore Your Wife? Query Answered by Famous Couples," *Lima News*, July 8, 1956.
8. Cochran and Brinley, *Jackie Cochran: The Autobiography*, p. 68.
9. Ibid., p. 61.
10. Ibid., p. 2.
11. Ibid., p. 3.
12. Cochran, *The Stars at Noon*, p. 84.
13. Ibid., p. 86.
14. "Description of Series," Jacqueline Cochran Collection Finding Aid, p. 1, Dwight D. Eisenhower Presidential Library, Abilene, KS.
15. Susan Ware, *Holding Their Own: American Women in the 1930s* (New York and London: Twayne, 1982), p. 25.
16. For a discussion on how society campaigned against women in the workforce, see William Henry Chafe, *The American Woman: Her Changing Social, Economic, and Political Roles, 1920–1970* (London and New York: Oxford University Press, 1972).
17. Lois Banner, *Women in Modern America: A Brief History* (San Diego: Harcourt Brace Jovanovich, 1984), pp. 85, 86.
18. Lois W. Banner, *American Beauty* (New York: Alfred A. Knopf, 1983), p. 272.
19. Ware, *Holding Their Own*, p. 75.
20. Jacqueline Cochran to Florence Wessels, Interview for the *Journal American*, Invitations, Stories, Broadcasts, Congratulations, 1938 (1), General Files #5, Jacqueline Cochran Collection, Dwight D. Eisenhower Presidential Library, Abilene, KS.
21. Cochran and Brinley, *Jackie Cochran: The Autobiography*, p. 119.
22. Ibid.
23. Ibid., p. 118.
24. Ibid., p. 119.
25. Ibid., p. 120.
26. Ibid., pp. 117, 118.
27. Jacqueline Cochran to unnamed interviewer, "Suggestion for Broadcast." Unidentified interview, Invitations, Stories, Broadcasts, Congratulations, 1938 (2), General Files #5, Jacqueline Cochran Collection, Dwight D. Eisenhower Presidential Library, Abilene, KS.
28. "Miss Cochran Buys Building on 56th St.," *New York Times*, September 16, 1945, VIII, p. 1:3.
29. "Top Officers of New Cosmetic Concern," *New York Times*, January 22, 1948, p. 40.
30. Maureen Joyce, "Jacqueline Cochran Odlum Dies at 73; Famous Aviatrix and Cosmetic Maker," *Washington Post*, August 10, 1980, B6.
31. Untitled Speeches about Amelia Earhart, Misc. Aviation Speeches (2), Speech Series, Box #1, Jacqueline Cochran Collection, Dwight D. Eisenhower Presidential Library, Abilene, KS.
32. Ibid.
33. Ware, *Still Missing*, p. 48.
34. Cochran, Jacqueline, "The Amelia I Knew," Unidentified Text, Radio Broadcasts 1946-1947-1948, General Files #1, Jacqueline Cochran Collection, Dwight D. Eisenhower Presidential Library, Abilene, KS.
35. Cochran and Brinley, *Jackie Cochran: The Autobiography*, p. 141.
36. Ibid., p. 135.
37. Ware, *Still Missing*, p. 86.
38. Mary S. Lovell, *The Sound of Wings: The Life of Amelia Earhart* (New York: St. Martin's Press, 1989), p. 12.
39. Ibid., p. 196.
40. Ibid.
41. Cochran, Untitled Speech about Amelia Earhart, Misc. Aviation Speeches, Eisenhower Presidential Library.
42. "No. 1 Woman Flier Jacqueline Cochran, Who Flew a Bomber to Britain, Advises American Women to Follow a British Example," *New York Times Magazine*, July 13, 1941, p. 11.
43. Ware, *Still Missing*, p. 94.
44. Ibid., p. 50.
45. Cochran and Brinley, *Jackie Cochran: The Autobiography*, p. 140.
46. Ibid.
47. Ibid., p. 141.
48. Jacqueline Cochran, Untitled Speech about Amelia Earhart, Misc. Aviation Speeches (2), Speech Series Box #1, Jacqueline Cochran collection, Dwight D. Eisenhower Presidential Library, Abilene, KS.
49. Ibid.
50. Cochran and Brinley, *Jackie Cochran: The Autobiography*, p. 141.
51. Cochran, *The Stars at Noon*, p. 91. and Cochran and Brinley, *Jackie Cochran: The Autobiography*, p. 142.
52. Lovell, *The Sound of Wings*, p. 297

Notes — Chapter 5

53. Cochran and Brinley, *Jackie Cochran: The Autobiography*, p. 142.

54. In 1938, Eleanor Roosevelt contacted Cochran and asked for specific data regarding Earhart's disappearance. Cochran then wrote to pilot Paul Mantz, who had flown with Earhart in the Bendix, and requested specific information to launch a new search. This White House connection may account for the popular theory that Amelia Earhart disappeared while on a spy mission for President Roosevelt.—Jacqueline Cochran to Paul Mantz, April 11, 1938, Paul Mantz (1938), General Files #5, Jacqueline Cochran Collection, Dwight D. Eisenhower Presidential Library, Abilene, KS.

55. Cochran and Brinley, *Jackie Cochran: The Autobiography*, p. 143.

56. Ibid.

57. Cochran, *The Stars at Noon*, p. 91.

58. Lovell, *The Sound of Wings*, p. 400.

59. Frankie Basch to Jacqueline Cochran, Transcript of Radio Broadcast Interview with Jacqueline Cochran, WMCA, New York, April 11, 1938, Talks and Speeches, 1938, General Files #5, Jacqueline Cochran Collection, Dwight D. Eisenhower Presidential Library, Abilene, KS.

60. Ibid.

61. Cochran, *The Stars at Noon*, p. 97.

Chapter 5

1. Cochran and Brinley *Jackie Cochran: The Autobiography*, pp. 165, 166.

2. For a brief description of the WAFs see, Winifred C. Cullis, "What British Women Are Doing in the War," Lecture delivered at Vassar College, April 30 and May 1, 1942; and Margaret Biddle, *The Women of England* (Boston: Houghton Mifflin, 1941).

3. "The Women's Auxiliary Territorial Service," Fact Sheet, ATA Miscellaneous, ATA #5, Air Transport Box #5, Jacqueline Cochran Collection, Dwight D. Eisenhower Presidential Library, Abilene, KS.

4. Jack Cassin-Scott, *Women at War, 1939–1945* (London: Osprey, 1980), p. 12.

5. Ibid.

6. Jacqueline Cochran to Eleanor Roosevelt, September 28, 1939, Aviation Correspondence August–December 1939 (2), General Files #6, Jacqueline Cochran Collection, Dwight D. Eisenhower Presidential Library, Abilene, KS.

7. Ibid.

8. Ibid.

9. Ibid.

10. Adelaide Handy, "Aviatrix Sees Advance in Woman's Cause Paid For in Loss of Homage to Femininity," *New York Times*, May 12, 1940, p. 5.

11. Jacqueline Cochran, "Women in Air Defense," Unidentified Speech, 2, Cochran, Jacqueline Documents, CC-344500-01, National Air and Space Museum, Washington, D.C.

12. Cochran, *The Stars at Noon*, p. 98.

13. Ibid, p. 101.

14. Ibid.

15. Ibid., p. 98.

16. Ibid.

17. Cochran and Brinley, *Jackie Cochran: The Autobiography*, p. 168.

18. Ibid., pp. 170, 171.

19. "Jacqueline Cochran Flies Bomber to Britain for Service with R.A.F.," *New York Times*, June 21, 1941, p. 5.

20. Cochran, *The Stars at Noon*, p. 101.

21. Elizabeth R. Valentine, "No. 1 Woman Flier Jacqueline Cochran, Who Flew a Bomber to Britain, Advises American Women to Follow a British Example," *New Times Magazine*, July 13, 1941, p. 11; and "U.S.A. Woman Pilot's Achievement: First to Fly Bomber Across Atlantic," *London Times*, June 21, 1941, p. 4.

22. "U.S.A. Women Pilot's Achievement," *London Times*, June 21, 1941, p. 4.

23. Lettice Curtis, *The Forgotten Pilots: A Story of the Air Transport Auxiliary 1939–1945* (Olney, Buckinghamshire: Nelson and Saunders, 1971), pp. 142, 143.

24. A.R. Glancy, Memoir, August 5, 1941, Bomber Flight England 1941 (3), WASP series, Box #1, Jacqueline Cochran Collection, Dwight D. Eisenhower Presidential Library, Abilene, KS.

25. "Jacqueline Cochran Flies Bomber to Britain for Service with R.A.F.," *New York Times*, June 21, 1941, p. 5.

26. Ibid.

27. Ibid.

28. Jacqueline Cochran, "Bombers as Bundles for Britain," Stories Written by Me—Jackie, Article Series, Box #1, 10, Jacqueline Cochran Collection, Dwight D.

Notes — Chapter 5

Eisenhower Presidential Library, Abilene, KS.
29. Ibid., pp. 5, 11, 12.
30. Ibid., 6.
31. Ibid., p. 5.
32. Ibid.
33. Ibid., p. 6.
34. Pauline Gower, "The ATA Girls," *Flying* 33 (August 1943), p. 30.
35. Cochran, "Bombers as Bundles for Britain," p. 9.
36. Ibid., p. 9.
37. Untitled Question and Answer Fact Sheet Concerning the ATA, from July 1, 1941 to July 1942, ATA (2), Air Transport Box #5, Jacqueline Cochran Collection, Dwight D. Eisenhower Presidential Library, Abilene, KS.
38. Jacqueline Cochran, "Women Fly in the War," *Salute* (British War Relief Society Magazine), fall 1941, p. 32, Cochran, Jacqueline Documents, Vertical Files, CC-344500-01, National Air and Space Museum, Washington, D.C.
39. Cochran, *The Stars at Noon*, pp. 105, 106; and "Miss Cochran a Guest: Flier Has Luncheon with the Roosevelts at Hyde Park," *New York Times*, July 3, 1941, p. 9.
40. Eleanor Roosevelt to General Watson, July 3, 1941, Official File 249, Aeronautics, Eleanor Roosevelt Collection, Franklin D. Roosevelt Library, Hyde Park, New York.
41. Cochran, *The Stars at Noon*, p. 107.
42. Granger, *On Final Approach*, p. 6.
43. Ibid.
44. Jacqueline Cochran to Major General B.K. Yount, Memorandum, February 10, 1943, 1, ATA (3), Air Transport Box #5, Jacqueline Cochran Collection, Dwight D. Eisenhower Presidential Library, Abilene, KS.
45. Ibid.
46. Jacqueline Cochran to Bebe (no other name given), "Stars and Stripes in Britain," Transcript, North American Transmission, ATA Miscellaneous, Air Transport Box #5, Jacqueline Cochran Collection, Dwight D. Eisenhower Presidential Library, Abilene, KS.
47. Ibid., p. 2.
48. D. Corbett to Jacqueline Cochran, Royal Air Force Delegation (British Air Commission), December 11, 1941, Original Letter from British Air Commission, Air Transport Box #5, Jacqueline Cochran Collection, Dwight D. Eisenhower Presidential Library, Abilene, KS.
49. Jacqueline Cochran, Final Drat of Telegram, January 23, 1942, Originals of Letters and Data Telegrams and Letters Sent January 23, 1942, Air Transport Box #5, Jacqueline Cochran Collection, Dwight D. Eisenhower Presidential Library, Abilene, KS.
50. Jacqueline Cochran to Bebe (no other name given), "Stars and Stripes in Britain," North American Transmission, ATA Miscellaneous, Air Transport Box #5, Jacqueline Cochran Collection, Dwight D. Eisenhower Presidential Library, Abilene, KS.
51. Sally Van Wagenen Keil, *Those Wonderful Women in Their Flying Machines: The Unknown Heroines of World War II* (New York: Four Directions Press, 1979), pp. 72, 73, 74, 77.
52. Cochran and Brinley, *Jackie Cochran: The Autobiography*, p. 190.
53. Curtis, *Forgotten Pilots*, p. 143.
54. Glenn Kerfoot, "Propeller Annie," *Aviation Quarterly* 5 (4th Quarter, 1979), p. 336.
55. Ernie Pyle, "Women Fliers," Newspaper Clipping, October 19, 1942, ATA Group Helen Richey, Air Transport Box #4, Jacqueline Cochran Collection, Dwight D. Eisenhower Presidential Library, Abilene, KS.
56. Opal Anderson, Rough Draft of An Article about American ATA Girls, ATA Group — Opal Anderson, Air Transport Box #4, Jacqueline Cochran Collection, Dwight D. Eisenhower Presidential Library, Abilene, KS.
57. Keil, *Those Wonderful Women in Their Flying Machines*, p. 101.
58. Ibid., p. 99.
59. Granger, *on Final Approach*, p. 18.
60. Curtis, *Forgotten Pilots*, p. 143.
61. Keil, *Those Wonderful Women in Their Flying Machines*, p. 97.
62. Gower, "The ATA Girls," p. 32.
63. "Ancient and Tattered: The Arduous Task of the Air Transport Auxiliary," *London Times*, January 29, 1941, p. 5.
64. *Fly Girls*, Documentary Video, 1999.
65. Gower, "The ATA Girls," *Flying*, p. 30; and "A.T.A.'s Third Birthday: 100,000 Aircraft Delivered," *London Times*, September 17, 1942, p. 2.

Notes — Chapter 6

66. Curtis, *Forgotten Pilots*, p. 143.
67. Cochran, *The Stars at Noon*, p. 111.
68. Ibid., pp. 113, 114.
69. Ibid., pp. 112, 113.
70. Michael S. Sherry, *The Rise of American Air Power: The Creation of Armageddon* (New Haven and London: Yale University Press, 1987), p. 119.
71. Glynne M. Jones to Jacqueline Cochran, Headquarters 8th Air Service Command Air Section, July 3, 1942, Study of Ferry Service June — July, WASP Series, Box #2, Jacqueline Cochran Collection, Dwight D. Eisenhower Presidential Library, Abilene, KS.
72. Granger, *On Final Approach*, p. 20.
73. Cochran, *The Stars at Noon*, p. 112.
74. Keil, *Those Wonderful Women in Their Flying Machines*, p. 83.
75. Cochran and Brinley, *Jackie Cochran: The Autobiography*, p. 192.
76. Ibid.

Chapter 6

1. Jacqueline Cochran, "Women in Air Defense," Unidentified Speech, p. 2, Cochran, Jacqueline Documents, CC-344500-01, National Air and Space Museum, Washington, D.C.
2. Anderson, *Wartime Women*, p. 4.
3. Rosalind Rosenberg *Divided Lives: American Women in the Twentieth Century* (New York: Hill and Wang, Noonday Press, 1992), p. 135.
4. Chafe, *The American Woman*, p. 195.
5. Treadwell, *The Women's Army Corps*, p. 15.
6. Campbell, *Women at War with America*, p. 20.
7. Ibid.
8. Campbell, *Women at War with America*, p. 36.
9. Milkmann, *Gender at Work*, p. 50.
10. Hartmann, *The Home Front and Beyond*, p. 45.
11. Keil, *Those Wonderful Women in Their Flying Machines*, p. 108.
12. "Wafs Leader Here to Start Flying Force," Newspaper Clipping, *Wilmington Morning News*, September 11, 1942 (no page number), Love, Nancy Harkness — Vertical Files, CL-802000-01, National Air and Space Museum, Washington, D.C.
13. Cochran, *The Stars at Noon*, pp. 117, 118.
14. Ibid., p. 118.
15. Ibid., p. 119.
16. Scharr, *Sisters in the Sky*, p. 3.
17. Cochran, *The Stars at Noon*, p. 118.
18. Ibid.
19. Ibid.
20. Untitled Interview with Henry Arnold, Eight Questions and Answers in Arnold's handwriting, Washington Correspondence 1945 (4), WASP series, Box #4, Jacqueline Cochran Collection, Dwight D. Eisenhower Presidential Library, Abilene, KS.
21. Ibid., p. 2.
22. Hap Arnold to John F. Curry, January 17, 1942, Correspondence with General Arnold and General Olds: Reforming group in U.S. (4), WASP series, Box #1, Jacqueline Cochran Collection, Eisenhower Presidential Library, Abilene, KS.
23. Jacqueline Cochran to Henry Arnold, letter delivered by hand by Miss Cochran to General Arnold, Correspondence with General Arnold and General Olds, Reforming Group in U.S. (4), WASP Series, Box #1, Jacqueline Cochran Collection, Dwight D. Eisenhower Presidential Library, Abilene, KS.
24. Ibid.
25. Jacqueline Cochran to Hap Arnold, Use of Women Pilots, September 11, 1942, Correspondence with General Arnold and General Olds: Reforming Group in U.S. (4), WASP Series, Box #1, Jacqueline Cochran Collection, Dwight D. Eisenhower Presidential Library, Abilene, KS.
26. Ibid.
27. Cochran to Arnold, September 11, 1942.
28. Jacqueline Cochran to Henry Arnold, Memo For General Arnold (Very Personal and Confidential), No Date, 5, ATC Rebuttal — Drafts, Working Papers (1), WASP Series, Box #10, Jacqueline Cochran Collection, Dwight D. Eisenhower Presidential Library, Abilene, KS.
29. Ibid.
30. Ibid.
31. Cochran, *The Stars at Noon*, p. 118.
32. Jacqueline Cochran to Robert Olds, Memorandum concerning the Organization of Women Ferry Pilots, no date, Correspon-

183

Notes — Chapter 6

dence with General Arnold and General Olds Reforming Groups in U.S. (2), WASP Series, Box #1, Jacqueline Cochran Collection, Dwight D. Eisenhower Presidential Library, Abilene, KS.

33. Granger, *On Final Approach*, p. 38.
34. Scharr, *Sisters in the Sky*, p. 13.
35. Verges, *On Silver Wings*, p. 65.
36. Cochran to Henry Arnold, Memo (Very Personal and Confidential), p. 6, Eisenhower Presidential Library, Abilene, KS.
37. Cochran, *The Stars at Noon*, p. 119.
38. Oral History Interview of Miss Jacqueline Cochran by Captain Robert S. Bartanowicz and Major John "Fred" Shiner, 11–12 March 1976, Typed Transcript, p. [44], K239.0512-940, in USAF Collection, AFHRA, Maxwell Air Force Base, AL.
39. "Air Ferry School Opens for Women," *New York Times*, November 24, 1942, p. 21.
40. Cochran and Brinley, *Jackie Cochran: The Autobiography*, p. 199.
41. Scharr, *The WASP*, p. 4.
42. Cochran, *The Stars at Noon*, p. 120.
43. Cochran and Brinley, *Jackie Cochran: The Autobiography*, p. 200.
44. *Women of Courage: The Story of the Women Pilots of World War II*, Documentary Video, K.M. Productions, Inc., 1993.
45. "Address By General H.H. Arnold, Commanding General, Army Air Forces, Before WASP Ceremony, Sweetwater, Texas," December 7, 1944, War Department Bureau of Public Relations, Scrapbook 1944 (2), Scrapbook Series, Box #19, Jacqueline Cochran Collection, Dwight D. Eisenhower Presidential Library, Abilene, KS.
46. Granger, *On Final Approach*, pp. 71, 73.
47. Marjorie Kumler, "They've Done It Again!," *Ladies Home Journal* (March 1944), p. 29.
48. Ibid.
49. Ibid.
50. Granger, *On Final Approach*, pp. 71, 72.
51. According to Cochran in *The Stars at Noon*, p. 120, the women were housed at nearby hotels with free bus service to and from the airport. This information conflicts with Granger's story in *On Final Approach*, which describes various types of housing and no transportation. Since Cochran was absent the first day and, as Granger was one of the members of 43-1, Granger's account is probably correct.
52. Granger, *On Final Approach*, p. 72.
53. Ibid.
54. Barbara Shelby, "The Fifinellas," *Flying* 33 (July 1943), p. 166.
55. Frank J. Taylor, "Our Women Warbirds," *Liberty*, January 29, 1944, p. 27.
56. "Angel Flight: Women's Airforce World War II," *Yankee Air Force* (June 1984), p. 13.
57. Granger, *On Final Approach*, p. 76.
58. Ibid.
59. Oral History Interview of Cochran by Capt. Bartanowicz and Major Shiner, March 11–12, 1976, p. 66.
60. Granger, *On Final Approach*, p. 79.
61. *Women of Courage*, Documentary Video, 1993.
62. Granger, *Final Approach*, p. 79.
63. Cochran, *The Stars at Noon*, p. 121.
64. Cochran and Brinley, *Jackie Cochran: The Autobiography*, p. 207.
65. Ibid.
66. Jacqueline Cochran to Colonel K.P. McNaughton, Memorandum, Subject: Women's Flying Training, December 15, 1942, 2, WAC Data (2), WASP Series, Box #5, Jacqueline Cochran Collection, Dwight D. Eisenhower Presidential Library, Abilene, KS.
67. Ibid.
68. Granger, *On Final Approach*, p. 80.
69. Cochran and Brinley, *Jackie Cochran: The Autobiography*, p. 201.
70. Cochran, *The Stars at Noon*, p. 120.
71. "23 Women Pilots Take Ferry Jobs," *Wichita Daily Times*, April 25, 1943, p. 6, Air Force Units, WASPs, Vertical Files, S1005100, National Air and Space Museum, Washington, D.C.
72. Granger, *On Final Approach*, p. 118.
73. Ibid. p. 124.
74. Louise Keller, "5 WAAF's in City Say Glamour Doesn't Count," *Detroit Free Press*, January 17, 1943, Love, Nancy Harkness, Documents, Vertical Files, CL-802000-10, National Air and Space Museum, Washington D.C.
75. Cornelia Fort, "At the Twilight's Last Gleaming," *Women's Home Companion* 70 (July 1943), p. 19.
76. "Attention Women's Shows," Basic News, Public Relations (7), WASP Series.

Notes — Chapter 7

77. "No Fuss and Feathers for WAF Squadron's Boss," Newspaper Clipping, September 13, 1942, Love, Nancy Harkness, Documents, Vertical Files, CL-802000-01, National Air and Space Museum, Washington, D.C.
78. Ibid.
79. Granger, *On Final Approach*, p. 124.
80. Cochran's choice of colors apparently inspired the novelist Janet Dailey. In 1984, Dailey published a fictional account of the WASPs entitled *Silver Wings: Santiago Blue*. Cochran, *The Stars at Noon*, p. 124.
81. Despite Cochran's ambitious nature, she suggested that women pilots be in one overall program, even if she wasn't in charge. In a memo to General Arnold she stated, "I would rather lose my identity with the women's flying project and see it go well, than have all the rank and power you could give me in a set-up that could produce only mediocre results by comparison." Granger, *On Final Approach*, p. 119.
82. Cochran and Brinley, *Jackie Cochran: The Autobiography*, p. 214.
83. "'WASP' Is New Title for AAF Woman Pilot," Press Release, August 20, 1943, Press Release (1), WASP Series, Box #13, Jacqueline Cochran Collection, Dwight D. Eisenhower Presidential Library, Abilene, KS.
84. Scharr, *The WASP*, p. 16.
85. "Coup for Cochran," *Newsweek*, XXXII (July 19, 1943), p. 41, Love, Nancy Harkness, Vertical Files, CL-802000-09, National Air and Space Museum, Washington D.C.
86. Ibid.
87. Interview with Nancy Harkness Love, Untitled Transcript, April 16, 1944, p. 3, Resume of Conversation Between Love and Col. McCormick (no other name given) of Ramspeck Committee, Love, Nancy Harkness, Documents, Vertical Files, CL-802000-03, National Air and Space Museum, Washington, D.C.
88. Ibid.
89. War Department, Bureau of Public Relations, Press Branch, "New Uniform, Insignia, Adopted for the WASPs," Press Release, November 17, 1943, Press Releases (1), WASP Series, Box #13, Jacqueline Cochran Collection, Dwight D. Eisenhower Presidential Library, Abilene, KS.
90. Jean Hascall Cole, *Women Pilots of World War II* (Salt Lake City: University of Utah Press, 1992), p. 15.
91. Ibid, p. 35.
92. Oral History Interview of Miss Cochran, by Captain Bartanowicz and Major John Shiner, March 11–12, 1976, p. 65.
93. "Notes on the B-26 Training (WASPS)," Information Sheet Created by the Army Air Forces, Publicity (2), WASP Series, Box #14, Jacqueline Cochran Collection, Dwight D. Eisenhower Presidential Library, Abilene, KS.
94. Ibid.
95. Ibid.
96. Cole, *Women Pilots of World War II*, p. 79.
97. Ibid., p. 78.

Chapter 7

1. Ann R. Johnson, "The WASP of World War II," *Aerospace Historian* (Summer–Fall, 1970), p. 76, S1005100 Air Force Units, WASPs, Vertical Files, National Air and Space Museum, Washington, D.C.
2. Cole, *Women Pilots of World War II*, p. 153.
3. Barbara Selby, "The Fifinellas," *Flying* 33 (July 1943), p. 76.
4. Army Air Forces Historical Studies, *Women Pilots with AAF 1941–1944*, p. 16.
5. William E. Hall to Julius H. Anberg, Memorandum Regarding the Ramspeck Committee — Investigation of the WASP, April 21, 1944, Central Decimal Files October 1942–May 1944, 333.55 Congressional Investigations, Record Group 18 (Army Air Forces), Records of the Army Air Forces, National Archives, College Park, Maryland.
6. "Avenger's WASPs," *Pegasus* (November, 1943), p. 7.
7. "Miss Cochran Put in High Air Post," *New York Times*, July 6, 1943, p. 18; and "Begins Direction of Women Pilots," *New York Times*, July 7, 1943, p. 16.
8. Army Air Force History Office, "Women Pilots with the AAF 1941–1944," p. 65.
9. Ibid., p. 52.
10. Ibid., p. 53.
11. Ibid.
12. Cochran quoted in the AAF Historical Office, "Women Pilots with the AAF," p. 53.

Notes — Chapter 7

13. Campbell, *Women at War with America*, p. 19.

14. News Release, Women Air Force Service Pilots, S1005100, Air Force Units, WASPs, Vertical Files, National Air and Space Museum, Washington, D.C.

15. *Women of Courage*, Documentary Video.

16. Frank J. Taylor, "Our Women Warbirds," *Liberty*, January 29, 1944, p. 72.

17. Ibid.

18. Cole, *Women Pilots of World War II*, p. 117.

19. *Women of Courage*, Documentary Video.

20. Ibid.

21. Historical Branch, Intelligence and Security Division, Headquarters, Air Transport Command, *Women Pilots in the Air Transport Command*, Volume 1, 149, 300.071, January–June 1948 — 300.101-5, November–December 1942, p. 10, in USAF Collection, AFHRA, Maxwell Air Force Base, Montgomery, AL.

22. Barbara E. Poole, "Requiem for the WASP," *Flying* 35 (December 1944), p. 56.

23. Keil, *Those Wonderful Women in Their Flying Machines*, pp. 260, 261, 262.

24. *Women of Courage*, Documentary Video.

25. Ibid.

26. Morris R. Nelson, "Women Pilots in Tow Target Squadrons," Air Corps Directive, July 10, 1943, Central Decimal Files October 1942-May 1944, 231.21 Women Pilots 1942–1943 o 231.27 Experts, Record Group 18 (Army Air Forces), Official Records of the Air Adjutant General, Headquarters Army Air Forces, File Pertaining to Women Pilots, National Archives, College Park, MD.

27. AAF History Office, "Women Pilots with the AAF 1941–1944," p. 67.

28. Charlotte Knight, "She Wears A Pair of Silver Wings," *Air Force* (January 1944), p. 49, WASP Vertical Files, National Air and Space Museum, Washington, D.C.

29. Cole, *Women Pilots of World War II*, p. 85.

30. *Women of Courage*, Documentary Video.

31. Cole, *Women Pilots of World War II*, p. 87.

32. Keil, *Those Wonderful Women in Their Flying Machines*, p. 228.

33. Winifred Wood, *We Were WASPs* (Coral Gables, FL: Glade House, 1945), p. 161.

34. Ibid., p. 162.

35. *Women of Courage*, Documentary Video.

36. War Department, "Women Pilots' Services Expanded," p. 264.

37. W.O. Senter to Commanding General, Army Air Forces, Subject: Pilot Weather Officers Available for Reassignment, October 27, 1943, p. 2, S1005100 Air Force Units, WASPs, Vertical Files, National Air and Space Museum, Washington, D.C.

38. Granger, *On Final Approach*, p. 262.

39. Ibid.

40. War Department: Bureau of Public Relations, "WASPs Establish New Records For Safety in Military Flying," Press Release, January 4, 1944, Press Release (1), WASP Series, Box #13, Jacqueline Cochran Collection, Eisenhower Presidential Library, Abilene, KS.

41. Cochran, Final Report, 1 June, 1945, p. 33.

42. John Stuart, "The WASP," 34 *Flying* (Jan., 1944), p. 148.

43. Joan Whelan related her story in Cole's, *Women Pilots of World War II*, p. 84.

44. Jacqueline Cochran to Commanding General, Army Air Forces, WASP Report, 1 June, 1945, pp. 31, 33, WASP Final Report, WASP Series, Box #12, Jacqueline Cochran Collection, Dwight D. Eisenhower Presidential Library, Abilene, KS.

45. *Women of Courage*, Documentary Video; and Fort, "At the Twilight's Last Gleaming," p. 19.

46. Cole, *Women Pilots of World War II*, p. 1.

47. Scharr, *Sisters in the Sky*, p. 559.

48. Stuart, "The WASPs," p. 148.

49. *Women of Courage*, Documentary Video.

50. Ibid.

51. Grange, *On Final Approach*, p. 108.

52. Cole, *Women Pilots of World War II*, p. 67.

53. Scharr, *Sisters in the Sky*, p. 570.

54. *Women of Courage*, Documentary Video.

55. AAF Historical Office, "Women Pilots with the AAF," p. 68.

56. *Women of Courage*, Documentary Video.

Notes — Chapter 7

57. "War Department: Women Pilots' Services Expanded," *American Aviation Daily* (October 23, 1943), p. 264, Public Relations (6), WASP Series, Box #7, Jacqueline Cochran Collection, Eisenhower Presidential Library, Abilene, KS.
58. *Women of Courage*, Documentary Video.
59. Keil, *Those Wonderful Women in Their Flying Machines*, p. 249.
60. Cochran quoted in AAF Historical Office, "Women Pilots in the AAF 1941–1944," p. 55.
61. Ibid.
62. Florence Miller, "The WAFs Fly a Mission," *Pegasus* (no date), 1005100, Air Force Units, WASPs, Vertical Files, National Air and Space Museum, Washington, D. C.
63. Charlotte Knight, "Women as Service Pilots," *Skyways* 3 (February 1944), p. 70.
64. Untitled Story, Public Relations, WASP, Collection of Correspondence, Memos, etc., Mar-Aug. 1943, 220.0721-9 1942–1943, vol. 4 — 220.072-12, March–August 1943, Vol. 26, in USAF Collection, AFHRA, Maxwell Air Force Base, Montgomery, AL.
65. Wood, *We Were WASPs*, p. 121.
66. *Women of Courage*, Documentary Video.
67. Cochran to the Assistant Chief of the Air Staff, Report on Recent Trip to Bases Utilizing WASPs, August 31, 1944, p. 2. and Jacqueline Cochran to Commanding General, Army Air Forces, Subject: Women Air Force Service Pilots, August 1, 1944, Scrapbook 1944 (2), Scrapbook Series, Box #19, Jacqueline Cochran Collection, Eisenhower Presidential Library, Abilene, KS.
68. Army Air Forces History Office Headquarters of the Army Air Forces, "Women Pilots with the AAF 1941–1944," Bound Report, AAF Historical Study #55, 72–73. Record Group 407, RG NVM, Reference Collection, Document No. 1081 to 1084, Records of the Adjutant General's Office, Box no. 218, National Archives, College Park, MD.
69. Ibid., p. 72.
70. Ibid.
71. Ibid.
72. Ibid.
73. Chafe, *The American Woman*, p. 152.
74. Charlotte Knight, "Our Women Pilots," *Air Force* (September 1943), p. 10, S1005100, Air Force Units, WASPs, Vertical Files, National Air and Space Museum, Washington, D.C.
75. Wood, *We Were WASPs*, p. 76.
76. John Stuart, "The WASP," *Flying* 34 (Jan., 1944), p. 73.
77. *History of the Women Air Force Service Pilots in the Second Air Force*, 6, WASP 432.01, 1 November — 31 December 1949, 432.073-1, September 1943 — May 1945, 11. In USAF Collection, AFHRA, Maxwell Air Force Base, Montgomery, AL.
78. Ibid., p. 7.
79. Jacqueline Cochran to the Assistant Chief of Air Staff, Memorandum, Subject: Report on Recent Trip to Bases Utilizing WASPs, August 31, 1944, Women Army Service Pilots, 324.5 WAC, June to September 1944 to WASP, Army Air Forces, Air and Adjutant General, Mail and Records Division, Unclassified Records Section, Decimal File June 1944–1946, National Archives, College Park, MD.
80. Ibid.
81. Ibid., p. 2.
82. Knight, "Women as Service Pilots," *Skyways* 3 (February 1944), p. 29.
83. Ibid., p. 70.
84. Stuart, "The WASP," *Flying* (January 1944), p. 73.
85. Ibid., p. 163.
86. Ibid., pp. 5, 7, 13.
87. Frank J. Taylor, "Our Women Warbirds," *Liberty*, January 29, 1944, p. 72.
88. Milkman, *Gender at Work*, p. 61.
89. Charlotte Knight, "She Wears a Pair of Silver Wings," *Air Force* (Jan. 1944), p. 51.
90. "Girl Pilots," *Life*, July 19, 1943, p. 73.
91. "Avenger's WASPs," *Pegasus* (November 1943), no page number, S1005100 Air Force Units, WASPs Vertical Files, National Air and Space Museum, Washington, D.C.
92. Ibid.
93. Knight, "She Wears a Pair of Silver Wings," *Air Force* (January 1944), p. 50.
94. "Avenger's WASPs," *Pegasus* (November 1943), p. 13.
95. "Women: Saved from Official Fate," *Time* 43, April 3, 1944, p. 63.
96. Scharr, *Sisters in the Sky*, p. 144.
97. Ibid.
98. Cole, *Women Pilots of World War II*, p. 107.

99. Ibid.
100. Ibid., p. 153.
101. *Women of Courage*, Documentary Video.
102. Ibid.
103. "Avenger's WASPs," *Pegasus* (November, 1943), p. 12, S1005100 Air Force Units, WASPs, Vertical Files, National Air and Space Museum, Washington, D.C.
104. Cole, *Women Pilots of World War II*, p. 138.
105. Bartanwicz and Shiner, "Oral History of Miss Jacqueline Cochran," p. 37.
106. Ibid.
107. Chafe, *The American Woman*, pp. 138, 139.
108. Cochran, Oral History Interview, p. 91.
109. Ibid.
110. Cochran, *The Stars at Noon*, p. 128.
111. "Women: Saved From Official Fate," *Time* 43, April 3, 1944, p. 63; "Girl Pilots," *Life*, 73; and "Avenger's WASPs," *Pegasus* (November 1943), p. 5.
112. "Avenger's WASPs," *Pegasus* (November 1943), p. 12.
113. *Women of Courage*, Documentary Video.
114. Knight, "Women as Service Pilots," p. 70.
115. Cole, *Women Pilots of World War II*, p. 155.

Chapter 8

1. Scharr, *Sisters in the Sky*, p. 705.
2. "Future Predicted for Women Fliers," *New York Times*, November 2, 1944, p. 16.
3. Ibid.
4. Jacqueline Cochran to Commanding General, Army Air Forces, Subject: WASP Report, June 1, 1945, WASP Final Report, WASP Series, Box #12, Jacqueline Cochran Collection, Dwight D. Eisenhower Presidential Library, Abilene, KS.
5. Henry H. Arnold to the Chief of Staff, Memorandum, Subject: Militarization of Women Pilots, January 17, 1944, Militarization (1), WASP Series, Box #12, Jacqueline Cochran Collection, Dwight D. Eisenhower Presidential Library, Abilene, KS.
6. Ibid.
7. "Backs Bill to Put WASPs in the AAF," *New York Times*, March 23, 1944, p. 22.
8. U.S. Congress, House, Committee on Military Affairs, *A Bill to Provide for the Appointment of Female Pilots and Aviation Cadets in the Air Forces of the Army, Hearing before the House Committee on Military Affairs on H.R. 4219*, 78th Cong., 2d session, March 22, 1944, p. 4.
9. Ibid.
10. "Consider Inquiry on Retaining WASPs," *New York Times*, March 15, 1944, p. 22.
11. "WASP Militarization Favored by Stimson," *New York Times*, May 5, 1944, p. 2.
12. "Would Halt Rise of Women Pilots," *New York Times*, June 6, 1944, p. 10.
13. U.S., Congress, House, *Concerning inquiries Made of Certain Proposals for the Expansion and Change in Civil Service Status of the WASP*, Remarks from the Committee on the Civil Service quoted in a report to the Committee of the Whole House, *Committee on the Civil Service House of Representatives*, House Report 1600, 78th Cong., 2d Session, 5 June 1944, Box RG 18 (Army Air Forces) Central Decimal Files, Oct. 1942-May 1944, 324.5 WACS to 324.5 WASPs, National Archives, College Park, MD.
14. "Women: Unnecessary and Undesirable?" *Time* 43, May 29, 1944, p. 66.
15. Jacqueline Cochran to Robert Ramsbeck (Ramspeck), Letter (ca. 1944), Militarization (3), WASP Series, Box #12, Jacqueline Cochran Collection, Eisenhower Presidential Library, Abilene, KS.
16. "Pictures Male Fliers Doing WASPs' Chores," The *New York Times*, June 20, 1944, p. 11.
17. Ibid.
18. U.S. Congress, House, Congressman O'Hara speaking against appointing female pilots and aviation cadets in the Air Forces of the Army, H.R. Res. 4219, 78th Cong., 2d session, June 13,1944 to August 24, 1944, *Congressional Record* 90:6413.
19. U.S. House, Congressman Izac speaking against H.R. 4219, 78th Cong., 2d session, p. 6415.
20. Ibid.
21. "House Defeats Bill to Put WASPs in Army," p. 19.
22. Ibid.

23. U.S. Congress, House, Congressman Costello speaking for H.R. 4219, *Congressional Record*, 78th Cong., 2d session, p. 6414.

24. Since the commanding officer of the Women's Army Corps held the rank of colonel, it would be logical and fair to assume that the commanding officer of a militarized women's pilot organization would hold the same rank.

25. U.S. Congress, House Congressman Izac speaking against H.R. 4219, *Congressional Record*, 78th Cong., 2d session, p. 6414.

26. Granger, *On Final Approach*, p. 345.

27. U.S. Congress, House, Concerning H.R. 4219, 78th Cong., 2d session, June 13, 1944 to September 8, 1944, p. 3561.

28. D'Ann Campbell, "Women in the Military," *Choice* 31 (September 8, 1993), p. 63.

29. "WASP Training Set to End," *New York Times*, June 27, 1944, p. 16.

30. Jacqueline Cochran to Commanding General, Army Air Forces, Subject: Women Airforce Service Pilots, August 1, 1944, Press Releases (2), WASP Series, Box #13, Jacqueline Cochran Collection, Eisenhower Presidential Library, Abilene, KS.

31. "Army Status Asked for Women Pilots," *New York Times*, August 8, 1944, p. 20.

32. Cochran to Commanding General, August 1, 1944.

33. Ibid.

34. "Director of Women Pilots Asks for Military Status For WASPs," Press Release, War Department, Bureau of Public Relations, August 8, 1944, 1, Overall Report, Draft, Working Papers (3), WASP Box #13, Jacqueline Cochran Collection, Dwight D. Eisenhower Presidential Library, Abilene, KS.

35. Keil, *Those Wonderful Women in Their Flying Machines*, p. 316.

36. Granger, *On Final Approach*, p. 394.

37. Merryman, *Clipped Wings: The Rise and Fall of the Women Airforce Service Pilots (WASPs) of World War II*, p. 110.

38. Scharr, *Sisters in the Sky*, p. 660.

39. As Scharr points out at the time of the August controversy, Cochran was delaying with the death of one of Floyd Odlum's sons, who she referred to as "one of our two sons." Her grief may have interfered with her judgment. See Scharr, *Sisters in the Sky*, p. 660.

40. B. Kimball Baker, "Uncle Sam's Flying Nieces," *Aviation* Quarterly 7 (Second Quarter, 1984), p. 262.

41. Arnold wrote a similar letter to Cochran stating that "the reduction in the flying training program and the changing war situation's bearing on availability and deployment of pilots make it evident that the WASP will soon become pilot material in excess of needs." Henry Arnold to Director of Women Pilots (Jacqueline Cochran), October 1, 1944, Henry Arnold Papers, Container 204, Reel 205, Frame 166, The Library of Congress, Washington, D.C.

42. Henry Arnold to Each Member of the WASP, Letter, October 1, 1944, Deactivation, WASP Series, Box #12, Jacqueline Cochran Collection, Dwight D. Eisenhower Presidential Library, Abilene, KS.

43. Cochran and Brinley, *Jackie Cochran: The Autobiography*, p. 214.

44. Ibid.

45. Although Cochran declined to protest, the directors of the General Federation of Women's Clubs unanimously adopted a resolution to protest the Congressional disbandment of the WASPs without granting them military status. "Aid to Youth Asked of 2,500,000 Women," *New York Times*, October 17, 1944, p. 17.

46. "Statement by Miss Jacqueline Cochran on Accomplishments of WASP Program," Press Release, War Department, Bureau of Public Relations, December 19, 1944, 1, Press Releases (3), WASP Series, Box #13, Jacqueline Cochran Collection, Dwight D. Eisenhower Presidential Library, Abilene, KS.

47. *Women of Courage*, Documentary Video.

48. Ibid.

49. Ibid.

50. Ibid.

51. B. Kimberly Baker, "Uncle Sam's Flying Nieces," *Aviation* Quarterly 7 (Second Quarter, 1984), p. 262.

52. *Women of Courage*, Documentary Video.

53. Cole, *Women Pilots of World War II*, p. 136.

54. Jan Churchill, *On Wings to War: Teresa James, Aviator* (Manhattan, KS: Sunflower University Press, 1992), p. 163.

55. Cole, *Women Pilots of World War II*, p. 137.
56. Ibid., p. 140.
57. Ibid.
58. Ibid., 138.
59. The clandestine atmosphere surrounding Cochran's postwar tour has fueled a lot of speculation, especially in connection with Amelia Earhart. After Cochran's trip, theories arose that she traveled to Japan to release Earhart, supposedly captured by the Japanese in 1937. In 1971, Cochran strongly denied that she brought Earhart out of Japan. Two Earhart biographers, Susan Ware and Mary Lovell, give the Japanese captivity theory scant attention, with no mention of a Cochran angle. See "Jacqueline Cochran: A Renaissance Woman of the 20th Century," *The Retired Officer* (September, 1971), p. 29; Susan Ware, *Still Missing*, pp. 226–227; and Mary Lovell, *The Sound of Wings*, pp. 323–333.
60. M.C. Devilbliss, *Women and Military Service: A History, Analysis, and Overview of Key Issues* (Maxwell Air Force Base, AL: Air University Press, 1990), p. 21.
61. Jeanne Holm, *Women in the Military: An Unfinished Revolution* (Novato, California: Presidio Press, 1993), p. 447.
62. Oral History Interview of Miss Jacqueline Cochran, March 11–12, 1976, p. 5, Maxwell Air Force Base, Montgomery, AL.
63. Ibid., p. 100.
64. *Fly Girls*, Documentary Video, 1999.
65. Scharr, *Sisters in the Sky*, p. 742.
66. "Says Goldwater: Women Pilots are Veterans," Provo, Utah *Herald*, October 24, 1977.
67. Doris Brinker Tanner, "We Also Served," *American History Illustrated* 20 (November 1985), p. 49.
68. Huma Khan, "First Female Aviation Pilots Get Gold Honor: 'I Never Thought It Would Happen,'" May 10, 2010, ABC News/Politics, http//abcnews.go.com/politics/women/air/service/pilots

Chapter 9

1. B. Kimberly Baker, "Uncle Sam's Flying Nieces," *The Aviation Quarterly* 7 (Second Quarter, 1984), p. 243.
2. Yeager and Janos, *Yeager: An Autobiography*, p. 275.
3. Ibid.
4. "Miss Cochran Sets Mark," *New York Times*, December 30, 1949, p. 7.
5. Philip H. Ableson, "Jet-Powered Flight," *Science*, Vol. 254, No. 5031 (October 25, 1991), p. 497.
6. Martin Van Creveld, *The Age of Airpower* (New York: Public Affairs, 2011), p. 192.
7. Abelson, "Jet-Powered Flight," p. 497.
8. Albert Speer, *Inside the Third Reich* (New York: Macmillan, 1970), p. 362.
9. "Jet Propelled Planes," *Science News Letter* (October 7, 1944), p. 227.
10. A.C. Monahan, "America's Jet Age," *Science News Letter* (August 9, 1952), p. 90 (inclusive: pp. 90–91).
11. H.H. Arnold, *Global Mission* (New York: Harper and Brothers, 1949), p. 242.
12. Wesley Frank Craven, James Lea Cate, eds. *Men and Planes*, Vol. 6 of *The Army Air Forces in World War II*, in seven volumes (Chicago: University of Chicago Press, 1955), p. 234.
13. Arnold, *Global Mission*, p. 533.
14. A.C. Monahan, "America's Jet Age," *Science News Letter* (August 9, 1952), p. 90 (inclusive: pp. 90–91).
15. Miss Graham (no other name given), "Interview with Miss Cochran for "American Weekly," November 12, 1948, Jacqueline Cochran—Articles #1, Jacqueline Cochran Collection, Dwight D. Eisenhower Presidential Library, Abilene, KS.
16. Ibid.
17. Enid Shomer, *Stars at Noon: Poems from the Life of Jacqueline Cochran* (Fayetteville: University of Arkansas Press, 2001), p. 50.
18. Yeager and Janos, *Yeager*, p. 268.
19. Ibid., p. 269.
20. Doris L. Rich, *Jackie Cochran: Pilot in the Fast Lane* (Gainesville: University of Florida Press, 2007), p. 176.
21. *Jackie Cochran: First Lady of Flight*, Video Documentary, 50 minutes, Global Science Productions, 2002.
22. Yeager and Janos, *Yeager*, p. 285.
23. *Jackie Cochran: First Lady of Flight*, Video Documentary.
24. Yeager and Janos, *Yeager*, p. 285.
25. "Miss Cochran Tops the Speed of Sound," *New York Times*, July 19, 1953, p. 32.

26. Yeager and Janos, *Yeager*, p. 285.

27. Jennifer E. Manning, Colleen J. Shogan, Susan Navarro Smelcer, *Women in the United States Congress: 1917–2011*, CRS Report for Congress, March 18, 2011, p. 4.

28. Drew Pearson, "Washington Merry-Go-Round," *San Mateo Times*, October 20, 1956, p. 18.

29. Shomer, *Stars at Noon*, p. 65.

30. Rich, *Jacqueline Cochran: Pilot in the Fast Lane*, p. 198.

31. "Cochran's Airborne Campaign, *Life*, May 7, 1956, p. 65.

32. "Jacqueline Cochran Odlum Visits in Bard-Winterhaven," *Sun*, October 12, 1956.

33. Rich, *Jackie Cochran: Pilot in the Fast Lane*, p. 196.

34. Shomer, *Stars at Noon*, pp. 66–67.

35. Ibid., p. 67.

36. Ibid.

37. Tom Patterson, "Triumph and Tragedy of Dalip Saund," *California Historian* (June 1992), www.pbs.org/rootsinthesand/dalip.

38. Drew Pearson, "Washington Merry-Go-Round," *San Mateo Times*, October 20, 1956, p. 18.

39. Patterson, "Triumph and Tragedy."

40. Transcript, Jacqueline Cochran Oral History Interview I, April 7, 1974, interview by Joe B. Frantz, Lyndon B. Johnson Presidential Library, Austin, TX.

41. "President Kennedy's Speech and America's Next Moonshot Moment," The National Aeronautics and Space Administration Website, www.nasa.gov.topics/history/features/kennedy.

42. Rich, *Pilot in the Fast Lane*, p. 206.

43. Frank Sis, "Top Aviatrix Foresees Space Women," Pacific *Stars and Stripes*, July 21, 1961.

44. "Aviatrix Sets Speed Record," Delaware County, PA, *Daily News*, August 26, 1961.

45. Sis, "Top Aviatrix Foresees Space Women," Pacific *Stars and Stripes*, July 21, 1961.

46. Tom Tiede, "Woman's World, But Space is Man's Domain," Racine *Sunday Bulletin*, September 7, 1969.

47. "Qualifications for Astronauts," House of Representatives Special Subcommittee on the Selection of Astronauts — Committee on Science and Astronauts, July 17, 1962, p. 38.

48. "Tiede, "Woman's World, But Space is Man's Domain."

49. Ibid.

50. Deborah G. Douglas, *American Women and Flight Since 1940* (Lexington: University of Kentucky Press, 2004), p. 152.

51. "Qualifications for Astronauts," House of Representatives Special Subcommittee on the Selection of Astronauts — Committee on Science and Astronautics," July 17, 1962, p. 31.

52. Margaret A. Weitekamp, "Lovelace's Woman in Space Program," National Aeronautics and Space Administration Website, www.nasa/gov.

53. "Aviatrix Cochran sets 2 Women's Speed Marks," Pacific *Stars and Stripes*, May 5, 1963.

54. Margo McLoone, *Women Explorer of the Air, Harriet Quimby, Bessie Coleman, Amelia Earhart, Beryl Markham, Jacqueline Cochran* (Mankato, MN: Capstone, 2000), pp. 42, 43.

55. Transcript Jacqueline Cochran Oral History Interview, I, April 7, 1974, interview by Joe B. Franz, Lyndon Johnson Presidential Library, Austin, TX.

56. Shomer, *Stars at Noon*, p. 65.

57. Ibid., p. 78.

58. Jacqueline Cochran, "I sat on the Doorstep of Heaven," *Parade*, May 16, 1954.

Bibliography

Ableson, Philip H. "Jet-Powered Flight." *Science*, Vol. 254, No. 5031 (October 25, 1991), p. 497.

Adams, Michael C.C. *The Best War Ever: America and World War II*. Baltimore and London: Johns Hopkins University Press, 1994.

"Address by General H.H. Arnold, Commanding General, Army Air Forces, Before WASP Ceremony, Sweetwater, Texas." December 7, 1944. War Department Bureau of Public Relations. Scrapbook 1944 (2), Scrapbook Series, Box #19, Jacqueline Cochran Collection, Dwight D. Eisenhower Presidential Library, Abilene, KS.

"Aid to Youth Asked of 2,500,000 Women." *New York Times*, October 17, 1944, p. 17.

"Aid Ferry School Opens for Women." *New York Times*, November 24, 1942, p. 21.

"Air and Space Museum, Biog." Louise McPhetridge Thaden (1989-0132), Vertical Files, CT-141000-01, National Air and Space Museum, Washington, D.C.

"Airplane Races Too Hazardous An Adventure for Women Pilots." *New York American*. Newspaper Clipping, August 26, 1929. Louise Thaden Vertical Files, CT-141000-02, National Air and Space Museum, Washington, D.C.

"An American Aviatrix Abroad: Miss Ruth Nichols Flies and Studies Aviation on Trip Around the World." *Aero Digest*, September 1925. Ruth Rowland Nichols Vertical Files, CN-06000-01. National Air and Space Museum, Washington, D.C.

The American Heritage Dictionary. Boston: Houghton Mifflin, 1982.

"Ancient and Tattered: The Arduous Task of the Air Transport Auxiliary." *London Times*, January 29, 1941, p. 5.

Anderson, Karen. *Wartime Women: Sex Roles, Family Relations, and the Status of Women During World War II*. Westport, CT: Greenwood Press, 1981.

Anderson, Opal. Rough Draft of an Article about American ATA Girls. ATA Group — Opa Anderson, Air Transport Auxiliary (ATA) Series, Box #4, Jacqueline Cochran Collection, Dwight D. Eisenhower Presidential Library, Abilene, KS.

"Angel Flight: Women's Air Force W.W.II." *Yankee Air Force* (June 1984), pp. 12, 13.

Armstrong, Harry G. *Principles and Practice of Aviation Medicine*. Baltimore: Williams and Wilkins, 1943.

Army Air Forces History Office Headquarters of the Army Air Forces. "Women Pilots with the AAF 1941-1944." Bound Report, AAF Historical Study #55, Record Group 407, RG NVM Reference Collection, Document No. 1081 to 1084, Records of the Adjutant General's Office, Box #218, National Archives College Park, MD.

"Army Status Asked for Women Pilots." *New York Times*, August 8, 1944, p. 20.

Arnold, H.H. "Air Power for Peace." *National Geographic Magazine* LXXXIX (February 1946), pp. 137-193.

Arnold, H.H. *Global Mission*. New York: Harper and Brothers, 1949, p. 242.

Arnold, Hap, to Jacqueline Cochran. September 19, 1941. Corresponding with General Arnold and General Olds: Reforming Group in U.S. (4). WASP Series. Box #1. Jacqueline Cochran Collection. Dwight D. Eisenhower Presidential Library, Abilene, KS.

Arnold, Henry, to Chief of Air Staff. Memorandum. Subject: Deactivation of WASP. October 1, 1944. Henry Arnold Papers.

Bibliography

Microfilm Reel 205. (Library of Congress, Washington, D.C.)

Arnold, Henry, to the Chief of Staff. Memorandum. Subject: Militarization of Women Pilots. January 17, 1944. Militarization (1), WASP Series, Box #12, Jacqueline Cochran Collection, Dwight D. Eisenhower Presidential Library, Abilene, KS.

Arnold, Henry, to Director of Women Pilots (Jacqueline Cochran). October 1, 1944. Henry Arnold Papers, Container 204, Reel 205, Frame 166, The Library of Congress, Washington, D.C.

Arnold, Henry, to Each Member of the WASP. Letter. October 1, 1944. Deactivation, WASP Series, Box #12, Jacqueline Cochran Collection, Dwight D. Eisenhower Presidential Library, Abilene, KS.

Arnold, Henry, to General Marshall. Memorandum. Subject: Incorporation of Women Civilian Pilots and Trainee into Army Air Forces. June 14, 1943. Sweetwater (3). WASP Series. Box #5. Jacqueline Cochran Collection. Dwight D. Eisenhower Presidential Library, Abilene, KS.

Arnold, Henry, to H.H. Burton. Letter. (No Date). Militarization (3). WASP Series. Box #12. Jacqueline Cochran Collection. Dwight D. Eisenhower Presidential Library, Abilene, KS.

Arnold, Henry, to T.C. Odom. Memorandum. Subject: Women Pilots. July 22, 1943. WAC Data (2). WASP Series. Box #5. Jacqueline Cochran Collection. Dwight D. Eisenhower Presidential Library, Abilene, KS.

"Attention Women's Shows." Basic News. Public Relations (7), WASP Series, Box #7, Jacqueline Cochran Collection, Dwight D. Eisenhower Presidential Library, Abilene, KS.

"Avenger's WASPs." *Pegasus* (November 1943), pp. 5–7, 13.

"Aviation Award Won By a Woman Flier." *New York Times*, February 19, 1938, p. 17.

"Aviation: Earth's Distance Shrivel as Melbourne Race Flyers Cut Air Travel from Britain to Australia to Two Days." *News-Week*, October 27, 1934.

"Aviatrix Cochran sets 2 Women's Speed Marks." *Pacific Stars and Stripes* (May 5, 1963).

"Aviatrix Sets Speed Record." Delaware County, PA. *Daily News* (August 26, 1961).

"Aviatrix Wrecks Plane." *New York Times*, June 3, 1947, p. 27.

"Award to Jacqueline Cochran." *New York Times*, June 3, 1947, p. 27.

"Awarded $12,500 Against Field of Ten Others." *Longview Texas News* Newspaper Clipping. September 4, 1938, No Page Number. Scrapbook 1938 (3). Scrapbook Series. Jacqueline Cochran Collection. Dwight D. Eisenhower Presidential Library, Abilene, KS.

"Backs Bill to Put WASPs in the AAF." *New York Times*, March 23, 1944, p. 22.

Bail, Robert E. "Record Set by Woman: Miss Cochran Made World Speed Mark in Plane Strange to Her." *New York Times*, September 26, 1937, p. 4.

Baker, B. Kimberly. "Uncle Sam's Flying Nieces." *The Aviation Quarterly* 7 (Second Quarter, 1984), p. 243.

Banner, Lois. *American Beauty*. New York: Alfred A. Knopf, 1983.

Banner, Lois. *Elizabeth Cady Stanton: A Radical for Women's Rights*. Boston: Little, Brown, 1980.

Banner, Lois. *Women in Modern America: A Brief History*. San Diego: Harcourt Brace Jovanovich, 1984.

Basch, Frankie, to Jacqueline Cochran. Transcript of Radio Broadcast Interview with Jacqueline Cochran. WMCA. New York. April 11, 1938. General File Series, Box #5, Jacqueline Cochran Collection, Dwight D. Eisenhower Presidential Library, Abilene, KS.

"Begins Direction of Women Pilots." *New York Times*, July 7, 1943, p. 16.

Bell, Harold. "Millville Incorporated For Less Than 13 Years Before Becoming Part of Greater Panama City." *Panama City News*, July 26, 1959.

Biddle, Margaret. *The Women of England*. Boston: Houghton Mifflin, 1941.

Bilstein, Roger E. *Flight in America: From the Wright Brothers to the Astronauts*. Baltimore and London: Johns Hopkins University Press, 1984.

Boyne, Walter J. *Beyond the Wild Blue: A History of the United States Air Force 1947–1997*. New York: St. Martin's Press, 1997.

Brett, George H. "The Air Force Struggle for Independence." *Air Power History* 13 (Fall 1996), pp. 22–29.

Brooks-Pazmany, Kathleen. *United States

Bibliography

Women in Aviation 1919–1925. Washington and London: Smithsonian Institution Press, 1991.

Brown, Dorothy. *Setting a Course: American Women in the 1920s*. Boston: Twayne, 1987.

Burton, H.H., to Henry Arnold. Letter. (No Date). Militarization (3). WASP Series. Box #12. Jacqueline Cochran Collection. Dwight D. Eisenhower Presidential Library, Abilene, KS. "Business and Finance." *Time*, August 19, 1935.

Cadogen, Mary. *Women with Wings: Female Flyers in Fact and Fiction*. Chicago: Academy Chicago, 1993.

Campbell, D'Ann. *Women at War with America: Private Lives in a Patriotic Era*. Cambridge and London: Harvard University Press, 1984.

Campbell, D'Ann. "Women in Uniforms: The World War II Experiment." *Military Affairs* 51 (July 1987), pp. 137–139. "Women in the Military." *Choice* 31 (September 1993), pp. 63–70.

Cassin-Scott, Jack. *Women at War 1939–45*. London: Osprey, 1980.

Chafe, William. *The American Woman: Her Changing Social, Economic, and Political Roles, 1920–1970*. London and New York: Oxford University Press, 1972.

Chant, Christopher. *Aviation: An Illustrated History*. London: Orbis, 1978.

Chun, Victor K. "The Origin of the WASPs." *American Aviation Historical Society* 14 (Winter 1969), pp. 259–262.

Churchill, Jan. *On Wings to War: Teresa James, Aviator*. Manhattan, KS: Sunflower University Press, 1992.

Cochran, Jackie. "My Flying." Stories Written by Me — Jackie (1), Articles Series, Box #1, Jacqueline Cochran Collection. Dwight D. Eisenhower Presidential Library, Abilene, KS.

Cochran, Jackie, and Maryann Bucknum Brinley. *Jackie Cochran: The Autobiography*. Toronto and New York: Rufus, 1987.

Cochran, Jacqueline. "The Amelia I Knew." Unidentified Text. Radio Broadcasts 1946-1947-1948, General Files, Box #1, Jacqueline Cochran Collection, Dwight D. Eisenhower Presidential Library, Abilene, KS.

Cochran, Jacqueline, to the Assistant Chief of Air Staff. Memorandum. Subject: Report on Recent Trip to Bases Utilizing WASPs, August 31, 1944. Women Army Service Pilots, 324.5 WAC. June to September 1944 to WASP, Army Air Forces, Air and Adjutant General, Mail and Records Division, National Archives, College Park, MD.

Cochran, Jacqueline, to Barney M. Giles. Memorandum. Subject: Militarization of Women Pilots through the WAC. July 26, 1943. Militarization (2). WASP Series. Box #5. Jacqueline Cochran Collection. Dwight D. Eisenhower Presidential Library, Abilene, KS.

Cochran, Jacqueline, to Bebe (no other name given). "Stars and Stripes in Britain." North American Transmission ATA Miscellaneous, Air Transport Auxiliary (ATA) Series, Box #5, Jacqueline Cochran Collection, Dwight D. Eisenhower Presidential Library, Abilene, KS.

Cochran, Jacqueline. "The Bendix Air Race." Miscellaneous Aviation Speeches (2), Speech Series, Box #1, Jacqueline Cochran Collection, Dwight D. Eisenhower Presidential Library, Abilene, KS.

Cochran, Jacqueline. "Bombers as Bundles for Britain." Report on Cochran's activities in Great Britain. Stories Written by Me — Jackie (1), Articles Series, Box #1, Jacqueline Cochran Collection, Dwight D. Eisenhower Presidential Library, Abilene, KS.

Cochran, Jacqueline, to Colonel K.P. McNaughton. Memorandum on Reasons for not Militarization on Women's Flying Training Program at the Present Time. December 15, 1942. WAC Data (2), WASP Series, Box #5, Jacqueline Cochran Collection, Dwight D. Eisenhower Presidential Library, Abilene, KS.

Cochran, Jacqueline, to Commanding General, Army Air Forces, Subject: WASP Report. June 1, 1945. WASP Final Report, WAASP Series, Box #12, Jacqueline Cochran Collection, Dwight D. Eisenhower Presidential Library, Abilene, KS.

Cochran, Jacqueline, to Commanding General, Army Air Forces, Subject Women Airforce Service Pilots. August 1, 1944. Press Releases (2). WASP Series. Box #13. Jacqueline Cochran Collection, Dwight D. Eisenhower Presidential Library, Abilene, KS.

Cochran, Jacqueline, to Eleanor Roosevelt. September 28, 1939. Aviation Correspondence August — December 1939 (2), Gen-

Bibliography

eral Files Services, Box #6, Jacqueline Cochran Collection, Dwight D. Eisenhower Presidential Library, Abilene, KS.

Cochran, Jacqueline. Final Draft of Telegram. January 23, 1942. Originals of Letters and Data Telegrams Sent January 23, 1942. Air Transport Auxiliary (ATA) Series, Box #5, Jacqueline Cochran Collection, Dwight D. Eisenhower Presidential Library, Abilene, KS.

Cochran, Jacqueline, to Florence Wessels. Interview for the *Journal American*. Invitations, Stories, Broadcasts, Congratulations, 1938 (1). General File Series, Box #5. Jacqueline Cochran Collection. Dwight D. Eisenhower Presidential Library, Abilene, KS.

Cochran, Jacqueline, to Frankie Basch. "Success Stories." WMCA Radio Program. April 12, 1938. New York City, New York. Transcript. Talks and Speeches, 1938 (2). General File Series. Box #5. Jacqueline Cochran Collection. Dwight D. Eisenhower Presidential Library, Abilene, KS.

Cochran, Jacqueline, to George M. Lewis. August 1, 1940. General Files #8. Collier Trophy 1940 (3), Jacqueline Cochran Collection, Dwight D. Eisenhower Presidential Library, Abilene, KS.

Cochran, Jacqueline, to Hap Arnold. Use of Women Pilots. September 11, 1942. Correspondence with General Arnold and General Olds: Reforming Group in U.S. (4), WASP Series, Box #1, Jacqueline Cochran Collection, Dwight D. Eisenhower Presidential Library, Abilene, KS.

Cochran, Jacqueline, to Henry Arnold. Letter delivered by hand by Miss Jacqueline Cochran to General Arnold. Correspondence with General Arnold and General Olds Reforming Group in U.S. (4). WASP Series. Box #1. Jacqueline Cochran Collection. Dwight D. Eisenhower Presidential Library, Abilene, KS.

Cochran, Jacqueline, to Henry Arnold. Memo for General Arnold. (Very Personal and Confidential). No Date. ATC Rebuttal — Drafts, Working Papers (1), WASP Series, Box #10, Jacqueline Cochran Collection, Dwight D. Eisenhower Presidential Library, Abilene, KS.

Cochran, Jacqueline. "I sat on the Doorstep of Heaven." *Parade* (May 16, 1954).

Cochran, Jacqueline, to Jane Sullivan. February 19, 1976. Jacqueline Cochran. Vertical Files. CC-344500-01. (National Air and Space Museum, Washington, D.C.).

Cochran, Jacqueline to Major General B.K. Yount. Memorandum. February 10, 1943, ATA (3), Air Transport Auxiliary (ATA) Series, Box #5, Jacqueline Cochran Collection, Dwight D. Eisenhower Presidential Library, Abilene, KS.

Cochran, Jacqueline, to Robert Olds. Memorandum Concerning the Organization of Women Ferry Pilots. Correspondence with General Arnold and General Olds Reforming Groups in U.S. (2), WASP Series, Box #1, Jacqueline Cochran Collection, Dwight D. Eisenhower Presidential Library, Abilene, KS.

Cochran, Jacqueline, to Robert Ramsbeck (Ramspeck). Letter (circa 1944). Militarization (3), WASP Series, Box #12, Jacqueline Cochran Collection, Dwight D. Eisenhower Presidential Library, Abilene, KS.

Cochran, Jacqueline, to Paul Mantz. April 11, 1938. Paul Mantz (1938). General File Series, Box #5, Jacqueline Cochran Collection, Dwight D. Eisenhower Presidential Library, Abilene, KS.

Cochran, Jacqueline. "I Reached the Stars the Hard Way." *Life*, August 16, 1954, p. 107.

Cochran, Jacqueline. *The Stars at Noon*. Boston and Toronto: Little, Brown, 1954.

Cochran, Jacqueline. "Talk on Unification." May 31, 1946. Armed Forces Unification Speech. Speech Series. Box #1. Jacqueline Cochran Collection. Dwight D. Eisenhower Presidential Library, Abilene, KS.

Cochran, Jacqueline, to Unnamed Interviewer. "Suggestions for Broadcast." Unidentified Interview. Invitations, Stories, Broadcasts, Congratulations, 1938 (2). General File Series, Box #5. Jacqueline Cochran Collection, Dwight D. Eisenhower Presidential Library, Abilene, KS.

Cochran, Jacqueline. Untitled Speech about Amelia Earhart. Misc. Aviation Speeches (2), Speech Series, Box #1, Jacqueline Cochran Collection, Dwight D. Eisenhower Presidential Library, Abilene, KS.

Cochran, Jacqueline. "Women in Air Defense." Unidentified Speech, 2. Cochran, Jacqueline Documents, CC-3445000-01, National Air and Space Museum, Washington, D.C.

Bibliography

Cochran, Jacqueline. "Women Fly in the War." *Salute* (British War Relief Society Magazine), Fall 1941, pp. 30–32.

"Cochran's Airborne Campaign." *Life* (May 7, 1956), p. 65.

Cole, Jean Hascall. *Women Pilots of World War II*. Salt Lake City: Utah University Press, 1992.

"Collier Trophy Presentation." Newspaper Clipping. National Aeronautic Association, 1941. General File Series. Box #9. Jacqueline Cochran Collection. Dwight D. Eisenhower Presidential Library, Abilene, KS.

Committee on the Civil Service House of Representatives. 78th Congress, 2nd Session. "Investigation of Civilian Employment: Concerning Inquiries Made of Certain Proposals For the Expansion and Change in Civil Service Status of WASPs. Report no. 1600. United States Government Printing Office: Washington, 1944.

WASP Bill — Pending Legislation. Records Group 18. Army Air Forces. Central Decimal Files. Oct. 1942 — May 1944. 324.5 WAACS to 324.5 WASPs. National Archives, College Park, MD.

Connelly, Sheryl. "Just Plain Great: Jackie Cochran had the Stuff Legend are Made of." Newspaper Clipping, July 23, 1987. Cochran, Jacqueline Documents, CC-344500-01. National Air and Space Museum, Washington, D.C.

Connolly, Vera L. "Daughters of the Sky." *The Delineator* 115 (August 1929), pp. 9, 81–83.

"Consider Inquiry on Retaining WASPs." *New York Times*, March 15, 1944, p. 2.

"Conversation Between Miss Cochran, Colonel Hobby and Colonel Carmichael." Telephone Conversation Transcript. June 25, 1943. Sweetwater (3). WASP Series. Box #5. Jacqueline Cochran Collection. Dwight D. Eisenhower Presidential Library, Abilene, KS.

Cook, Blanche Weisen. *Eleanor Roosevelt*. New York: Viking, 1992.

Corn, Joseph J. *The Winged Gospel: America's Romance with Aviation, 1900–1950*. New York and Oxford: Oxford University Press, 1983.

Corbett, D., to Jacqueline Cochran. Royal Air Force Delegation (British Air Commission), December 11, 1941. Original Letter from British Air Commission, Air Transport Auxiliary (ATA) Series, Box #5, Jacqueline Cochran Collection, Dwight D. Eisenhower Presidential Library, Abilene, KS.

"Coup for Cochran." *Newsweek*, July 19, 1943, p. 41.

Craven, Wesley Frank, and James Lea Cate, eds. *Men and Planes*, Vol. 6 of *The Army Air Forces in World War II*, in seven volumes (Chicago: University of Chicago Press, 1955), p. 234.

Cullis, Winifred C. *What British Women are Doing in the War*. Lecture Delivered at Vassar College, April 30 and May 1, 1942.

Curtis, Lettice. *The Forgotten Pilots: A Story of the Air Transport Auxiliary 1939–1945*. Olney, Buckinghamshire: Nelson and Saunders, 1971.

"Dale Fulton Sets World Air Record." *New York Times*, September 1, 1946, p. 36.

Dailey, Janet. *Silver Wings: Santiago Blue*. New York: Pocket Books, 1984.

Davis, Allen F. *American Heroine: The Legend of Jane Addams*. New York: Oxford University Press, 1973.

de Leeuw, Hendrick. *Conquest of the Air: The History and Future of Aviation*. New York: Vantage Press, 1960.

De Seversky, Alexander P. *Victory Through Air Power*. New York: Simon & Schuster, 1942.

"Decrease is Noted in Private Flying." *New York Times*, August 7, 1948, p. 28.

"Description of Series." Finding Aid, Jacqueline Cochran Collection, Dwight D. Eisenhower Presidential Library, Abilene, KS.

Devilbliss, M.C. *Women and Military Service: A History, Analysis, and Overview of Key Issues*. Maxwell Air Force Base, AL: Air University Press, 1990.

"Director of Women Pilots Asks for Military Status for WASPs." Press Release. War Department. Bureau of Public Relations, August 8, 1944. Overall Report, Draft, Working Papers (3), WASP Series, Box #13, Jacqueline Cochran Collection, Dwight D. Eisenhower Presidential Library, Abilene, KS.

Dodd, Dorothy, to James Boyd. May 2, 1947. Products-Agriculture-Lumber-Sawmill. Clipping Files. (State Library of Florida, Tallahassee, FL.)

Bibliography

Douglas, Deborah G. *American Women and Flight Since 1940* (Lexington: University of Kentucky Press, 2004), p. 152.

Douglas, Deborah G. *United States Women in Aviation 1940–1985*. Washington and London: Smithsonian Institution Press, 1991.

Dwiggins, Don. *They Flew the Bendix Race: The History of the Competition for the Bendix Trophy*. Philadelphia and New York: J.B. Lippincott, 1965.

Earhart, Amelia. *The Fun of It: Random Records of My Own Flying and of Women in Aviation*. Chicago: Academy Chicago, 1932.

Evans, Sara M. *Born for Liberty: A History of Women in America*. New York: Free Press, 1989.

"Fair Nearly Ties Saturday Record." *New York Times*, September 15, 1940, p. 47.

Fass, Paula S. *The Damned and the Beautiful: American Youth in the 1920s*. Oxford: Oxford University Press, 1977.

"Feminists Stirred Over Woman Flier." *New York Times*, November 8, 1935, p. 25.

Fly Girls: Female Pilots in World War II, Documentary Video, originally aired on KET-TV on May 24, 1999.

Flynn, Sister Margaret. Daughters of Charity, Evansville, Indiana. Telephone Interview, December 21, 1995.

Fort, Cornelia. "At the Twilight's Last Gleaming." *Women's Home Companion*, July 1943, p. 19.

Francis, Devon. "Jackie Cochran Gets What She Wants and How." *Ashville Times*. Newspaper Clipping. September 25, 1938, No Page Number, Scrapbook 1938(3), Scrapbook Series, Box #19, Jacqueline Cochran Collection, Dwight D. Eisenhower Presidential Library, Abilene, KS.

"Frank Fuller, the Man to Beat in the Bendix." Newspaper Clipping. September 2, 1938, No Page Number, Scrapbook 1938 (3), Scrapbook Series, Box #9, Jacqueline Cochran Collection, Dwight D. Eisenhower Presidential Library, Abilene, KS.

Frantz, Joe B. Transcript, Jacqueline Cochran Oral History Interview I, April 7, 1974. Lyndon B. Johnson Presidential Library, Austin, TX.

Frisbee, John L. *Makers of the United States Air Force*. Office of Air Force History: United States Air Force, Washington, D.C., 1987.

Furman, Bess. "Army Status Asked for Women Pilots." *New York Times*, August 8, 1944, p. 20.

"Future Predicted for Women Fliers." *New York Times*, November 2, 1944, p. 16.

"F.W. Fuller Breaks Bendix Race Mark." *New York Times*, September 4, 1937, p. 7.

General Headquarters United States Army Air Forces, Pacific. "Invitational Travel Orders," August 25, 1945, Jacqueline Cochran Vertical Files. CC-344500-02. (National Air and Space Museum, Washington, D.C.)

George, Harold L., to Commanding General, Army Air Forces. Militarization of Women Ferry Pilots. November 30, 1942. WASP Data (2). WASP Series. Box #5. Jacqueline Cochran Collection. Dwight D. Eisenhower Presidential Library, Abilene, KS.

Germais (?), Joe, to Bill (no other name given). Letter entitled "An Astute Observation." June 28, 1994. Jacqueline Cochran Vertical Files. CC-344500-02. National Air and Space Museum, Washington, D.C.

"Girl Pilots." *Time*, July 19, 1943, pp. 73–78.

"Girl Wins Bet by Earning Pilot's License During Her Vacation of Three Weeks." *New York World Telegram*. August 18, 1932. No page number. Scrapbook 1932–1937, Scrapbook Series, Box #19, Jacqueline Cochran Collection, Dwight D. Eisenhower Presidential Library, Abilene, KS.

Glancy, A.R. Personal Memoir. August 5, 1941. Bomber Flight England 1941 (3), WASP Series, Box #1, Jacqueline Cochran Collection, Dwight D. Eisenhower Presidential Library, Abilene, KS.

Glines, Carroll V. *Roscoe Turner: Aviation's Master Showman*. Washington and London: Smithsonian Institution Press, 1995.

Gower, Pauline Commander, "The ATA Girls." *Flying*. 33 (August, 1943), pp. 30–32.

Graham, Frederick. "Mantz Roars 435 Miles an Hour to Win Bendix Trophy Air Race." *New York Times*, August 31, 1946, p. 1.

Graham, Miss (no other name given). "Interview with Miss Cochran for "American Weekly," November 12, 1948, Jacqueline Cochran – Articles #1, Jacqueline Cochran Collection, Dwight D. Eisenhower Presidential Library, Abilene, KS.

Granger, Byrd Howell. *On Final Approach: The Women Airforce Service Pilots of W.W.II*. Scottsdale, AZ: Falconer, 1991.

Guiterrez, Gail M. "The Sting of Discrimi-

Bibliography

nation — Women Airforce Service Pilots (WASP). *Journal of the West* XXXV (January 1996), pp. 15–23.

Hall, Jacqueline Dowd, et al. *Like A Family: The Making of a Southern Cotton Mill World*. New York and London: W.W. Norton, 1987.

Hall, William E., to Julius H. Amberg. Memorandum Regarding the Ramspeck Committee — Investigation of the WASP. April 21, 1944. Central Decimal Files October 1942 — May 1944, 333.55 Congressional Investigations. Record Group 18 (Army Air Forces), Records of the Army Air Forces, National Archives, College Park, MD.

Hallion, Richard P. *Test pilots: The Frontiersmen of Flight*. Washington, D.C. and London: Smithsonian Institution Press, 1988.

Handy, Adelaide. "Aviatrix Sees Advance in Woman's Cause Paid for Loss of Homage to Femininity." *New York Times*, May 12, 1940, p. 5.

"Happiness, Horror Mingle, Women Derbyists Fly On." Newspaper Clipping. August 25, 1929, No page number, Louise McPhertridge Thaden (1989-0132), Vertical Files, CT-141000-02, National Air and Space Museum, Washington, D.C.

Hartmann, Susan M. *The Home Front and Beyond: American Women in the 1940s*. Boston: Twayne, 1982.

Headquarters United States Army Strategic Air Forces. "Letter Orders." August 22, 1945. Jacqueline Cochran Vertical Files. CC-344500-02. National Air and Space Museum, Washington, D.C.

Historical Branch Intelligence and Security Division, Headquarters, Air Transport Command. *History of the Air Transport Command*. (1946). 188, 301.04.9. 1941–1942. 301.073-3. 15 December 1944, 15 January 1945. USAF Collection. AFHRA. (Maxell Air Force Base, Montgomery, AL.)

Historical Branch, Intelligence and Security Division, headquarters, Air Transport Command. *Women Pilots in the Air Transport Command*. Volume 1. 300.071 January–June 1948 300.101-5 November–December 1942, USAF Collection, AFHRA. (Maxwell Air Force Base, Montgomery, Alabama.)

History of the Women Air Force Service Pilots in the Second Air Force. WASP 432.01, 1 November — 31 December 1949, 432.073 — 1 September 1943 — May 1945. USAF Collection. AFHRA. (Maxwell Air Force Base, Montgomery, AL.)

Holm, Jeanne. *Women in the Military: An Unfinished Revolution*. Novato, CA: Presidio Press, 1982.

"House Defeats Bill to Put WASPs in Army." *New York Times*, June 22, 1944, p. 19.

Hoyt, Edwin P. *Angels of Death: Goering's Luftwaffe*. New York: Tom Dorsey, 1994.

"Information Sheet on Jacqueline Cochran." Talks and Speeches 1937. General File Series. Box #4. Jacqueline Cochran Collection. Dwight D. Eisenhower Presidential Library, Abilene, KS.

Interview with Glen Dishman, Colonel, United States Air Force (Ret.), by Rhonda L. Smith. March 14, 1997.

Interview with Jacqueline Cochran, by Florence Wessels. Interview for the *Journal American*. Invitations, Stories, Broadcasts, Congratulations, 1938 (1), General Files, Box #5, Jacqueline Cochran Collection, Dwight D. Eisenhower Presidential Library, Abilene, KS.

Interview with Nancy Love Harkness. Untitled Transcript. April 16, 1944. Résumé of Conversation Between Mrs. Love and Col. McCormick, Love, Nancy Harkness, Documents, Vertical Files, CL-802000-03, National Air and Space Museum, Washington, D.C.

"Interview — Miss Graham." *American Weekly*. November 12, 1948. Rough Draft of an Article. Article Series, Box #1, Jacqueline Cochran Collection, Dwight D. Eisenhower Presidential Library, Abilene, KS.

"J Cochran Wins Bendix Race, Los Angeles to Cleveland." *New York Times*, September 4, 1938, p. 1.

Jackie Cochran: First Lady of Flight, Video Documentary, 50 minutes, Global Science Productions, 2002.

"Jacqueline Cochran." Unidentified Information Sheet. Talks and Speeches 1938 (12), General File Series, Box #5, Jacqueline Cochran Collection, Dwight D. Eisenhower Presidential Library, Abilene, KS.

"Jacqueline Cochran: Achieving Heights in Flying and Cosmetics," Fact Sheet. Cochran, Jacqueline Documents, Vertical Files, CC-344500-01, National Air and Space Museum, Washington, D.C.

Bibliography

"Jacqueline Cochran — Biographical." Information Sheet. Publicity (3). WASP Series. Box #14. Jacqueline Cochran Collection. Dwight D. Eisenhower Presidential Library, Abilene, KS.

"Jacqueline Cochran: Breaking Aviation Records Is Only One of Her Many Activities," *U.S. Air Services*, Loose Article, January, 1938, Cochran, Jacqueline, Documents, Vertical 76 Files, CC-344500-01, National Air and Space Museum, Washington, D.C.

"Jacqueline Cochran Flies Bomber to Britain for Service with R.A.F." *New York Times*, June 21, 1941, p. 5.

"Jacqueline Cochran Gets Trophy." *New York Times*, April 5, 1938, p. 19.

"Jacqueline Cochran: A Renaissance Woman for the 20th Century." *The Retired Officer*. (September, 1971), pp. 28–32. Jacqueline Cochran Vertical Files, M1 USAF Museum, Wright Patterson Air Force Base, Dayton, OH.

"Jacqueline Cochran Odlum Visits in Bard-Winterhaven." *Sun* (October 12, 1956).

"Jacqueline Cochran Takes First Money in Bendix Air Race Over 10 Male Pilots." *Santa Rosa California Republican*. Newspaper Clipping. September 3, 1938, No Page Number, Scrapbook 1938 (3), Scrapbook Series, Box #19, Jacqueline Cochran Collection, Dwight D. Eisenhower Presidential Library, Abilene, KS.

"Jacqueline Cochran Wins Bendix Derby." *Dallas Times-Herald*. Newspaper Clipping. September 4, 1938, No Page Number, Scrapbook 1938 (3), Scrapbook Series, Box #19, Jacqueline Cochran Collection, Dwight D. Eisenhower Presidential Library, Abilene, KS.

"Jet Propelled Planes." *Science News Letter* (October 7, 1944), p. 227.

Jones, Glynne M., to Jacqueline Cochran. Headquarters 8th Air Service Command Air Section, July 3, 1942, Study of Ferry Service June-July, WASP Series, Box #2, Jacqueline Cochran Collection, Dwight D. Eisenhower Presidential Library, Abilene, KS.

Johnson, Ann R. "The WASP of World War II." *Aerospace Historian*. Clipping. (Summer-Fall, 1970), pp. 76–82, S1005100 Air Force Units, Vertical Files — WASPs, National Air and Space Museum, Washington, D.C.

Joyce, Maureen. "Jacqueline Cochran Odlum Dies at 73; Famous Aviatrix and Cosmetic Maker." *Washington Post*, August 10, 1980, B6.

Keegan, John. *The Battle for History: Refighting World War II*. New York: Vintage, 1995.

Keegan, John. *The Second World War*. New York: Penguin, 1990.

Keil, Sally Van Wagenen. *Those Wonderful Women in Their Flying Machines: The Unknown Heroines of World War II*. New York: Four Directions Press, 1979.

Keller, Louise. "5 WAAF's in City Say Glamour Doesn't Count." *Detroit Free Press*, January 17, 1943. Love, Nancy Harkness, Documents, Vertical Files, CL-802000-10, National Air And Space Museum, Washington, D.C.

Knight, Charlotte. "Our Women Pilots." Clipping. *Air Force* (September, 1943), pp.10–12, S1005100 Air Force Units, Vertical Files — WASPs, National Air and Space Museum, Washington, D.C. "She Wears A Pair of Silver Wings." *Air Force*. Newspaper Clipping. (January, 1944), pp. 49, 50–51. Vertical Files — WASPs, National Air and Space Museum, Washington, D.C. "Women as Service Pilots." *Skyways* 3 (February 1944), pp. 28–29, 70, 74.

Kraus, Walter F., to Commanding General, Army Air Forces, Washington, D.C. Subject: Absorption of Army Air Forces Women Pilots into the Women's Army Auxiliary Corps. December 19, 1942. WAC Data (2). WASP Series. Box #5. Jacqueline Cochran Collection. Dwight D. Eisenhower Presidential Library, Abilene, KS.

Kumler, Marjorie. "They've Done It Again!" *Ladies Home Journal*, March 1944, pp. 28, 29, 167–169.

"Laura Ingalls' Desire to Fly Amused Pilots." *Washington Herald*, September 13, 1935. Newspaper Clipping. No Page Number. Laura A. Ingalls, Vertical Files, CI-020020-01 National Air and Space Museum, Washington, D.C.

Lemons, J. Stanley. *The Women Citizen: Social Feminism in the 1920s*. Urbana, Chicago and London: University of Illinois Press, 1973.

Lopez, Donald S. *Aviation: A Smithsonian Guide*. New York: Macmillan, 1995.

Lovell, Mary S. *The Sound of Wings: The Life of Amelia Earhart.* New York: St. Martin's Press, 1989.

"The MacRobertson International Air Races: England — Australia." Conditions. London Race General (9), Annual Files Series, Box #2, Jacqueline Cochran Collection, Dwight D. Eisenhower Presidential Library, Abilene, KS.

Manning, Jennifer E., Colleen J. Shogan, Susan Navarro Smelcer. *Women in the United States Congress: 1917–2011.* CRS Report for Congress, March 18, 2011, p. 4.

McLoone, Margo. *Women Explorer of the Air, Harriet Quimby, Bessie Coleman, Amelia Earhart, Beryl Markham, Jacqueline Cochran.* Mankato, MN: Capstone, 2000, pp. 42, 43.

Mendenhall, Charles A., and Tom Murphy. *The Gee Bee Racers: A Legacy of Speed.* North Branch, MN: Specialty Press, 1979.

Merryman, Molly. *Clipped Wings: The Rise and Fall of the Women Airforce Service Pilots (WASPs) of World War II.* New York and London: New York University Press, 1998.

Milkman, Ruth. *Gender at Work: The Dynamics of Job Segregation by Sex During World War II.* Urbana and Chicago: University of Illinois Press, 1987.

Miller, Florence. "The WAFs Fly a Mission." *Pegasus.* Newspaper Clipping. S1005100. Air Force Units. Vertical Files — WASPs, National Air and Space Museum, Washington, D.C.

"Miss Cochran Buys Building on 56th St." *New York Times,* September 16, 1945, section VIII, p. 3.

"Miss Cochran First at Bendix, Also." *New York Times,* September 4, 1938, p. 1.

"Miss Cochran Gets Trophy." *New York Times,* June 16, 1939, p. 14.

"Miss Cochran a Guest: Flier Has Luncheon with the Roosevelts at Hyde Park." *New York Times,* July 3, 1941, p. 9.

"Miss Cochran Put in High Air Post." *New York Times,* July 6, 1943, p. 18.

"Miss Cochran Sets Mark." *New York Times,* December 30, 1949, p. 7.

"Miss Cochran Sets Woman's Air Mark." *New York Times,* September 21, 1937, p. 29.

"Miss Cochran 33,000 Feet Up." *New York Times,* March 25, 1939, p. 17.

"Miss Cochran Tops the Speed of Sound." *New York Times* (July 19, 1953), p. 32.

"Miss Laura Ingalls: First American Woman to Fly the Andes." *US Air Services: Feature Aeronautical Magazine Commercial and Military.* Newspaper Clipping. June 1934. No Page Number. Laura A. Ingalls, Vertical Files, CI-020020-01, National Air and Space Museum, Washington, D.C.

Mitchell, Broadus. *The Rise of the Cotton Mills in the South.* New York: Da Capo Press, 1968.

Monahan, A.C. "America's Jet Age." *Science News Letter* (August 9, 1952), p. 90 (inclusive: pp. 90–91).

Montgomery, David. *The Fall of the House of Labor: The Workplace, The State, and American Labor Activism, 1865–1925.* Cambridge University Press: Cambridge and New York, 1989.

Morris, Lloyd and Smith, Kendall. *Ceiling Unlimited: The Story of American Aviation from Kitty Hawk to Supersonics.* New York: Macmillan, 1953.

"Mrs. Odlum Gets Nevada Divorce." *New York Times,* October 8, 1935, p. 19.

Nelson, Morris R. "Women Pilots in Tow Target Squadrons." Air Corps Directive. July 10, 1943. Central Decimal Files October 1942 — May 1944. 231.21 Women Pilots to 1942–1943 to 231.27. Experts, Record Group 18 (Army Air Forces), Official Records of the Air Adjutant General, Headquarters Army Air Forces. File Pertaining to Women Pilots, National Archives, College Park, MD. News Release. "Women Air Force Service Pilots." S1005100, Air Force Units, Vertical Files — WASPs, National Air and Space Museum, Washington, D.C.

Nichols, Ruth Rowland. Untitled Biographical Information. Vertical Files. CN-06000-01. National Air and Space Museum, Washington, D.C.

Nicholson, Mary, to Mr. Horner. News Story Draft on Jacqueline Cochran. August 18, 1937. Talks and Speeches 1937. General File Series. Box #4. Jacqueline Cochran Collection. Dwight D. Eisenhower Presidential Library, Abilene, KS.

Bibliography

"No Fuss and Feathers for WAF Squadron's Boss." Newspaper Clipping. September 13, 1942. Love, Nancy Harkness, Documents, Vertical Files, CL-802000-01, National Air and Space Museum, Washington, D.C.

"No. 1 Woman Flier Jacqueline Cochran, Who Flew a Bomber to Britain, Advises American Women to Follow a British Example." *New York Times Magazine*, July 13, 1941, p. 11.

Norton, Mary Beth, ed. *Major Problems in American Women's History*. Lexington, MA: D.C. Heath, 1996.

"Notes on B-26 Training (WASPs)." Information Sheet Created by the Army Air Forces. Publicity (2), WASP Series, Box #14, Jacqueline Cochran Collection, Dwight D. Eisenhower Presidential Library, Abilene, KS.

Oakes, Claudia M. *United States Women in Aviation 1930–1939*. Washington and London: Smithsonian Institution Press, 1991.

Odlum, Jacqueline Cochran, to Eva H. Thompson. July 4, 1944. Letter Concerning Adoption. WASP Series. Box #14. Jacqueline Cochran Collection. Dwight D. Eisenhower Presidential Library, Abilene, KS.

Odom, T.C., to Henry Arnold. "Report of Investigation Regarding Proper Use of Women Pilots in the AAF." August 55, 1943. Militarization (1). WASP Series. Box #12. Jacqueline Cochran Collection. Dwight D. Eisenhower Presidential Library, Abilene, KS.

Oral History Interview of Miss Jacqueline Cochran, by Captain Robert S. Bartanowicz and Major John "Fred" Shiner. 11–12 March, 1976. Typed Transcript. K239.0512-940. USAF Collection, AFHRA. (Maxwell Air Force Base, Montgomery, AL.)

Oral History Interview of Stuart Symington, by Joe B. Frantz. Oral History Transcript. October 6, 1976. AC 81-17. (Lyndon B. Johnson Library, Austin, TX).

Patterson, Tom. "Triumph and Tragedy of Dalip Saund." *California Historian* (June 1992), www.pbs.org/rootsinthesand/dalip.

Pearson, Drew. "Washington Merry-Go-Round." *San Mateo Times* (October 20, 1956), p. 18.

"Piece of Metal Could Be Clue to Earhart." *Lexington Herald-Leader*, December 7, 1996, p. A4.

Planck, Charles E. *Women with Wings*. New York and London: Harper and Brothers, 1942.

Polk, R.L. and Company. *Florida Gazetteer and Business Directory 1911–1912*. Jacksonville: R.L. Polk.

Poole, Barbara A. "Requiem for the WASP." *Flying* 35 (December 1944), pp. 55–56, 146, 148.

"President Kennedy's Speech and America's Next Moonshot Moment." The National Aeronautics and Space Administration Website, www.nasa.gov.topics/history/features/kennedy.

"Prexy Prattle: Deep in the Heart of Texas." WASP Newsletter. (Postwar, No Date.) Air Force Units. Vertical Files — WASPs S1005100. (National Air and Space Museum, Washington, D.C.)

"Propeller Annie." *Aviation*. 5 (Fourth Quarter, 1979), 318–337.

Pyle, Ernie. "Women Fliers." Newspaper Clipping. December 19, 1942. ATA Group Helen Richey, Air Transport Auxiliary (ATA) Series, Box #4, Jacqueline Cochran Collection, Dwight D. Eisenhower Presidential Library, Abilene, KS.

"Qualifications for Astronauts." House of Representatives Special Subcommittee on the Selection of Astronauts — Committee on Science and Astronauts (July 17, 1962), p. 38.

"Records Shattered at Miami Meet." *The Bee Hive*. Newspaper Clipping. December 1937. No Page Number. Jacqueline Cochran, Vertical Files, CC-344500-02, National Air and Space Museum, Washington, D.C.

"Report on Question of Militarization of Women Pilots." Militarization (1). WASP Series. Box #12. Jacqueline Cochran Collection. Dwight D. Eisenhower Presidential Library, Abilene, KS.

Rich, Doris L. *Jackie Cochran: Pilot in the Fast Lane*. Gainesville: University of Florida Press, 2007, p. 176.

Riley, Glenda. *Inventing the American Woman: A Perspective on Women's History 1865 to the Present*. Arlington Heights, Ill.: Harlan Davidson, 1986.

Roosevelt, Eleanor, to General Watson. July 3, 1941. Official File 249, Aeronautics,

Bibliography

Eleanor Roosevelt Collection, Franklin D. Roosevelt Library, Hyde Park, New York.

Rose, Elihu. "The Court-Martial of Billy Mitchell." *The Quarterly Journal of Military History — Special Issues: Air Power* (Spring 1996), pp. 16–25.

Rosenberg, Rosalind. *Divided Lives: American Women in the Twentieth Century.* New York: Hill and Wang, The Noonday Press, 1992.

St. Johns, Adela Rogers. "Will Aviation End Wars?" Newspaper Clipping. October 2, 1938. Pp. 1–3. Jacqueline Cochran Vertical Files. CC-344500-02. (National Air and Space Museum, Washington, D.C.)

Scharr, Adela Riek. *The WASP.* Vol. II: *Sisters in the Sky.* St. Louis, MO: Patrice Press, 1988.

"Scott and Black Win, Flying to Melbourne in 71 Hours; Dutch Still Second; 2 Killed." *New York Times,* October 23, 1934, pp. 1, 3.

Selby, Barbara. "The Fifinellas." *Flying* 33 (July 1943), pp. 76, 78, 166–167.

Senter, W.O., to Commanding General, Army Air Forces. Subject: Pilot Weather Officers Available for Reassignment. October 27, 1943. S1005100. Air Force Units, Vertical Files — WASPs, National Air and Space Museum, Washington, D.C.

Sherry, Michael S. *The Rise of American Air Power: The Creation of Armageddon.* New Haven and London: Yale University Press, 1987.

Shomer, Enid. *Stars at Noon: Poems from the Life of Jacqueline Cochran.* Fayetteville: University of Arkansas Press, 2001, p. 50.

Sis, Frank. "Top Aviatrix Foresees Space Women." Pacific *Stars and Stripes* (July 21, 1961).

"Southwest Pacific Area, Personnel Air Shipment, Priority Clearance." Travel Form Regarding Jacqueline Cochran. August 25, 1945. Jacqueline Cochran Vertical Files. CC-344500-02. (National Air and Space Museum, Washington, D.C.)

Speer, Albert. *Inside the Third Reich.* New York: Macmillan, 1970, p. 362.

"Statement by Miss Jacqueline Cochran on Accomplishments of WASP Program." Press Release. War Department. Bureau of Public Relations. December 19, 1944. Press Releases (3), WASP Series, Box #13,

Jacqueline Cochran Collection, Dwight D. Eisenhower Presidential Library, Abilene, KS.

Stuart, John. "The WASPs." *Flying* 34 (January, 1944), pp. 73–74, 148, 163.

Stugardi, Ed. St. Providence Hospital. Montgomery, AL, Telephone Interview, 21 December, 1995.

"Suggestion for Broadcast." Unidentified Interview with Jacqueline Cochran. Invitations, Stories, Broadcasts, Congratulations, 1938 (2), General Files, Box #5, Jacqueline Cochran Collection, Dwight D. Eisenhower Presidential Library, Abilene, KS.

Tanner, Doris Brinker. "We Also Served." *American History Illustrated* 20 (November 1985), pp. 12–21, 47–49.

Taylor, Frank J. "Our Women Warbirds." *Liberty* (Jan. 29, 1944), pp. 26–27, 72.

Thaden, Louise. *High, Wide and Frightened.* New York: Stackpole and Sons, 1938.

Tiede, Tom. "Woman's World, but Space Is Man's Domain." *Racine Sunday Bulletin* (September 7, 1969).

"Top Officers of New Cosmetics Concern." *New York Times,* January 22, 1948, p. 40.

Treadwell, Mattie E. *United States Army in World War II: Special Studies, The Women's Army Corps.* Washington, D.C.: Department of the Army, 1954.

"23 Women Pilots Take Ferry Jobs." *Wichita Daily Times,* April 25, 1943, p. 6.

"Two American Flyers Jailed by Persians." *New York Times,* October 25, 1934, p. 7.

U.S. Congress, House, Committee on Military Affairs. *A Bill to Provide for the Appointment of Female Pilots and Aviation Cadets in the Air Forces of the Army, Hearings Before the House Committee on Military Affairs on H.R. 4219.* 78th Cong., 2nd Sess., March 22, 1944.

U.S. Congress, House. Concerning H.R. 4219. 78th Cong., 2nd Sess., 13 June, 1944 to 8 September, 1944.

U.S. Congress, House. *Concerning Inquiries Made of Certain Proposals for the Expansion and Change of Civil Service Status on the WASP.* Remarks from the Committee on the Civil Service quoted in a report to the Committee of the Whole House. *Committee on the Civil Service House of Representations.* House Report 1600. 78th Cong., 2nd Sess., 5 June 1944. Box RG 18 (Army Air

Bibliography

Forces) Central Decimal Files, Oct. 1942 – May 1944, 324.5 WACS to 324.5 WASPs, National Archives, College Park, MD.

U.S. Congress, House. Congressman Costello Speaking for H.R. 4219. *Congressional Record*. 78th Cong., 2nd Sess., 13 June, 1944 to 8 September, 1944.

U.S. House. Congressman Izac Speaking Against H.R. 4129. 78th Cong., 2nd Sess., 13 June, 1944 to 8 September, 1944.

"U.S.A. Woman Pilot's Achievement: First to Fly Bomber Across Atlantic." *London Times*, June 21, 1941, p. 4.

"United States Air Force Emerges From New Defense Organization." *Aviation Week* 47 (September 29, 1947), pp. 9, 10.

"Universal, Monday 22 [October] 1934." Press Release. London Flight General (8), Annual Series, Box #2, Jacqueline Cochran Collection, Dwight D. Eisenhower Presidential Library, Abilene, KS.

Untitled Story. Public Relations. WASP. Collection of Correspondence, Memos, etc. Mar-Aug. 1943, 220.0721-9 1942–1943, Vol. 4 – 220.072-12, Mr-Aug. 1943, Vol. 26. USAF Collection, AFHRA, Maxell Air Force Base, Montgomery, Alabama.

Untitled Interview Between Jacqueline Cochran and Mr. Uttal. National Air Races 1938 Miscellaneous, General File Series, Box #5, Jacqueline Cochran Collection, Dwight D. Eisenhower Presidential Library, Abilene, KS.

Untitled Interview with Henry Arnold. Eight Questions and Answers in Arnold's Handwriting. Washington Correspondence 1945 (4), WASP Series, Box #4, Jacqueline Cochran Collection, Dwight D. Eisenhower Presidential Library, Abilene, KS.

Untitled Interview with Henry Arnold. Typed Answers. Washington Correspondence 1945 (4), WASP Series, Box #4, Jacqueline Cochran Collection, Dwight D. Eisenhower Presidential Library, Abilene, KS.

Untitled Interview with Jacqueline Cochran. Talks and Speeches 1937. General File Series, Box #4. Jacqueline Cochran Collection. Dwight D. Eisenhower Presidential Library, Abilene, KS.

Untitled Question and Answer Fact Sheet Concerning the ATA From July 1, 1941 to July 1, 1942. ATA (2), Air Transport Auxiliary (ATA) Series, Box #5, Jacqueline Cochran Collection, Dwight D. Eisenhower Presidential Library, Abilene, KS.

Valentine, Elizabeth R. "No. 1 Woman Flier Jacqueline Cochran, Who Flew A Bomber to Britain, Advises American Women to Follow a British Example." *New York Times Magazine*, July 13, 1941, p. 11.

Van Creveld, Martin. *The Age of Airpower*. New York: Public Affairs, 2011, p. 192.

Verges, Marianne. *On Silver Wings: The Women Airforce Service Pilots of World War II 1942–1944*. New York: Ballantine, 1991.

"Wafs Leader Here to Start Flying Force." Newspaper Clipping. Wilmington *Morning News*. September 11, 1942. Love, Nancy Harkness – Vertical Files, CL-802000-01, National Air and Space Museum, Washington, D.C.

War Department, Bureau of Public Relations. "Award Distinguished Service Medal to Miss Jacqueline Cochran." March 1, 1945, WASP Series, Box #7, Jacqueline Cochran Collection, Dwight D. Eisenhower Presidential Library, Abilene, KS.

War Department: Bureau of Public Relations. "WASP Establish New Records for Safety in Military Flying." Press Release. January 4, 1944. Press Release (1). WASP Series. Box #13. Presidential Library, Abilene, KS.

War Department, Bureau of Public Relations, Press Branch. "New Uniform, Insignia, Adopted For the WASPs." Press Release. November 17, 1943. Press Releases (1), WASP Series. Box #13, Jacqueline Cochran Collection, Dwight D. Eisenhower Presidential Library, Abilene, KS.

War Department: Women Pilot's Services Expanded." *American Aviation Daily*. Clipping. (October 23, 1943), p. 264. Public Relations (6), WASP Series, Box #7, Jacqueline Cochran Collection, Dwight D. Eisenhower Presidential Library, Abilene, KS.

Ware, Susan. *Holding Their Own: American Women in the 1930s*. New York and London: Twayne, 1982.

Ware, Susan. *Still Missing: Amelia Earhart and the Search for Modern Feminism*. New York and London: W.W. Norton, 1993.

"Washington Merry Go Round." Rough Draft from Fort Worth Texas *Star-Telegram* August 6, 1944. Press Releases and Etc. (1).

Bibliography

WASP Series, Box #13, Jacqueline Cochran Collection, Dwight D. Eisenhower Presidential Library, Abilene, KS.

"WASP Militarization Favored by Stimson." *New York Times*, May 5, 1944, p. 2.

"'WASP' Is New Title for AAF Woman Pilots." Press Release. August 20, 1943. Press Release (1), WASP Series, Box #13, Jacqueline Cochran Collection, Dwight D. Eisenhower Presidential Library, Abilene, KS.

Weitekamp, Margaret A. "Lovelace's Woman in Space Program." National Aeronautics and Space Administration Website, www.nasa/gov.

White, M.G., to the Commanding General, Army Air Forces. Memorandum. Subject: Incorporation of Women Civilian Pilots into the Army Air Forces. June 20, 1943. Militarization (2). WASP Series. Box #5. Jacqueline Cochran Collection. Dwight D. Eisenhower Presidential Library, Abilene, KS.

Woloch, Nancy. *Women and the American Experience: A Concise History*. New York: McGraw-Hill, 1996.

"Woman Victor in Bendix Air Race; Beat Eight Men." *Rock Island Ill Argus*. Newspaper Clipping. September 3, 1938. No page number, Scrapbook 1938 (3), Scrapbook Series, Box #19, Jacqueline Cochran Collection, Dwight D. Eisenhower Presidential Library, Abilene, KS.

Women of Courage: The Story of the Women Pilots of World War II. Documentary Video. K. M. Productions, 1993.

"Women: Home by Christmas." *Time*, October 16, 1944, p. 68.

"Women: Saved from Official Fate." *Time*, April 3, 1944, p. 63.

"Women: Unnecessary and Undesirable?" *Time*, May 29, 1944, p. 66.

"The Women's Auxiliary Air Force. The WAFFS." Fact Sheet. ATA Miscellaneous, ATA #5, Air Transport Auxiliary (ATA) Series, Box #5, Jacqueline Cochran Collection, Dwight D. Eisenhower Presidential Library, Abilene, KS.

"Women's Progress in Air is Shown by Naming of 3 as 'Marking Pilots.'" Newspaper Clipping. October 5, 1935. Louise McPhertidge Thaden, (1989-0132), Vertical Files, CT-141000-03, National Air and Space Museum, Washington, D.C.

"Women's Work." Program from "Florida Crossroads." Video #714. February 23, 1995. Florida Public Television.

Wood, Winifred. *We Were WASPs*. Coral Gables, Florida: Glade House, 1945.

"Would Halt Rise of Women Pilots." *New York Times*, June 6, 1944, p. 10.

Yeager, Chuck, and Leo Janos. *Chuck Yeager: An Autobiography*. New York: Bantam, 1985.

Index

A-24 104
Adams, Amy 151
Adams, Ruth 127
"air circuses" 24
Air Force (magazine) 112
"air mindedness" 22–23
Air Transport Command (ATC) 73, 85, 94–95, 163
Allison, Rhea Hurrle 144
Almay 57
Amelia (film) 151, 161
American Ferry Command 73
American General Dynamics Corporation 136
Antoine's 58
Arden, Elizabeth 57
Armistead, George 44
Armstrong, Harry 38
Army Air Corps 40, 44, 51, 65, 79–80, 83, 98–99, 115, 124
Army Air Forces 92, 111, 117, 119–120, 131; 8th 78, 84, 87; 319th Flying Training Detachment (AAFFTD) 86
Army Flying Field 155
Army Nurse Corps 82
Arnold, Bruce 131
Arnold, Robert 3
Arnold, Henry "Hap" 68–69, 73–74, 78, 84–86, 88, 92, 94–96, 99, 118–119, 121, 123–125, 129, 134, 155, 156, 163
Arthritis Foundation 167
Ascani, Colonel Fred 138
AT-6 88, 103, 106
Atlantic Charter Conference 165
Atlas Corporation 167
Atwood, Harry N. 155
Auriel, Jacqueline 135, 146
Avenger Field 94, 96
Aviation Hall of Fame 149, 135

B-4 bags 102
B-17 86, 88
B-24, Bomber 72, 106

B-26 96–97
B-29 131
B-34 108
barnstorming 24
Baumgartner, Ann 135
Beechcraft 43–44
Bell, Alexander Graham 21
Bell Aircraft Company 134
Bell XP-59 *Aircomet* 134
Bellanca 28–92 44
Bendix, Vincent 42, 62, 158
Bendix Air Race 2, 24, 35, 38–39, 42–47, 50, 52–54, 56, 133, 148, 153, 159, 166
Bendix Aviation Company 159
Bendix Helicopter Corporation 159
Bendix Trophy 153
Berry, Anne 128
Boeing 21
Boeing DC-1 22
Boeing DC-3 22
Boeing 247 22
Boothby, Walter 38
Boyd, General Albert 136
Boylan, Margaret 92
Bradley, Omar N. 165
British Air Commission 74
British Air Transport Auxiliary (ATA) 68, 72, 74–77, 80, 83, 87, 94
British Air Transport Auxiliary Service 66
British Ferry Command 68–69
BT-13 88
Bureau of Air Commerce 163
Burton, Harold 122

Cagle, Myrtle T. 144
Canadair 135
Caraway, Hattie 138
Carlisle, Stanford Grafton 70
Chapin, Emily 75, 76
Churchill, Winston 165
Civil Aeronautics Administration (CAA) 119
Civil Air Patrol (CAP) 119, 128

Index

Clairol 57
Clewis, Kay 110
Clifford Burke Harmon International Trophies 2
Coanda, Henri 133
Cobb, Geraldine "Jerrie" 144–145
Cochran-Odlum Ranch 60
Cole, Jean Hascall 96, 113
Coleman Army Base 110
Collier Trophy Committee 35, 46, 68
Conner, Lillian 87
Constant, Max 44
Cordova, Frank 44
Costello, Bill 120
Costello, John 119, 122
Craft, Anne 128
Craigie, Brig. Gen. Laurence C. 135
Crosson, Marvel 49
Curry, John F. 85
Curtis, Lettice 70, 75–77
Curtiss, Glenn 21, 25, 158

Deaton, Leoti 89–92
Demoe, Babette 105
Denison House 160
Dietrich, Jan and Marion 144
Disney, Walt 90
Distinguished Flying Cross 157
Distinguished Service Medal 157, 162
Doolittle, Major James J. 159
Douglas (company) 21
Douglas, Donald 21

Earhart, Amelia 1, 4, 20, 26–29, 36, 45–49, 51, 53–54, 56, 59–64, 151–153, 159, 160, 166
Ederle, Gertrude 20
Edwards Air Force Base, CA. 136–137
Eisenhower, Dwight D. 56, 129, 135, 139–142, 153
Eldridge, Fred 140, 141

F-86 134, 136, 138
F-86 Sabre jets 134, 136
F-104 Starfighter 148
Fédération Aéronautique Internationale 37
Felton, Rebecca Latimer 138
Ferry Command 84, 86–87, 100, 102, 106, 108, 163
Ferry Division 87, 100, 101, 106
"Fifinella" 90
Flickinger, Donald 143
Flowing Velvet 59, 147
Flying (magazine) 106–107, 111
Fort, Cornelia 93, 106–107
Fox Movietone News 76

Frank, A.H. 84
Frank Phillips Trophy Race 47
French Air Metal 37
Friendship 160
Fuller, Frank 38, 43–45
Funk, Mary Wallace 144

Gagarin, Yuri 142
Gee Bee Racer 35, 39, 41, 44, 46–47, 149
Gelback Lee 44
General Electric I-A 134
General William E. Mitchell Memorial Plaque 46
George, Harold 84–85
Gillies, Betty 110
Gilmore 23
Glancy, A.R. 70–71
Gloster Meteor 133
Golbinec, Leona 96, 127
Goldwater, Barry 130
Gorlick, Sara 144
Gowen Field 110
Gower, Pauline 72
Granger, Byrd Howell 93, 104, 122, 124, 126
Grevenberg, M.E. 32

Hadley, Ross 44
Hammond, Paul 60
Harmon Trophy 46
Harper, R.W. 101
Harriman, W. Averell 71
Harris, Arthur T. 74, 78
Hart, Jane 144
Hawker "Hurricane" (British) 75
Hawkins, Sadie 104
Haynes Motor Company 159
Henderson, Clifford W. 42, 47–48
Hichney, John 44
Hill, Lister 122
Hixon, Jean F. 144
Hobby, Oveta Culp 82, 85, 92, 122, 124, 161, 162
Hooper, Mary 76
Hopkins, John Jay 135
House Bill 4219 123
House Military Affairs Committee 119
Houston Airfield 91
Houston Post 161, 162
Howard Hughes Airport 89
Howland Island 62, 160
Hughes, Howard 23, 56

Imperial Russian Naval Academy 167
Ingalls, Laura 26, 45
Inter City Aviation 163
Irimescu, Radu 51

Index

Itasca 63–64
Izac, Edward 121–122

Jacqueline Cochran Air Show 153
Jacqueline Cochran Cosmetics 50, 54, 58–59, 146
Jacqueline Cochran, Inc. 59
James, Teresa 103, 127, 130
Johnny Livingston Air Circus 34, 36
Johnson, Claudia "Lady Bird" 166
Johnson, Lyndon B. 141, 147, 153, 166

Kai-shek, Chiang 41
Keil, Mary Ellen 103–104
Kelly, Florence Prag 138
Kelly Airmail Act 23
Kennedy, John F. 142
Kindig, G.C. "Brownie" Brown 91, 104
Klingensmith, Florence 47
Knight, Charlotte 112
Knight, Clayton 68
Korean War (1950–53) 166
Kumler, Marjorie 89

Lackawanna Railroad 158
Lady Drummon-Hay Trophy 37
LaJotte, Charles 44
Laraway, Fran 114, 128
Law, Ruth 20, 25–26, 65
Legion of Merit Award 149
Lemay, Helen 56
Leonard, Royal 41
Leverton, Iren 144
Lewelleyn, Husky 31
Liberty (magazine) 102, 111
Lindbergh, Anne Morrow 26, 47
Lindbergh, Charles 20–23, 26, 47
Lockheed Electra 60, 63, 160
Lockheed Jet F-104, Starfighter 146, 148
Lockheed Hudson 68, 70
Lockheed Orion 44
London, Barbara Erickson 126
London Times 70, 76
Lord Beaverbrook 69
Love, Nancy Harkness 26, 47, 84, 86, 93–94, 100, 103, 108, 162, 163, 164
Love, Robert 26, 84, 163
Lovelace, Randolph 38
Lovelace, Dr. Robert 143
Lovelace Foundation for Medical Education and Research 143
Lovett, Robert 73

Mackay Trophy 157
MacRobertson International Air Race 33–35, 37, 40–42, 51, 72
Magid, Elizabeth MacKethan 107

Mantz, Paul 44–45
Mari 63
Marshall, Anne Dailey 102, 126
Marshall, George C. 82, 118, 164, 165
Marshall, Ted 33
Martin, Madge Rutherford 130
Maru 63
May, Andrew 121
Maybelline 57
McNaughton, Kenneth 92
McQuarrie, Hortense 55, 166
Me-262 133, 134, 136
Medal of Merit 169
Mercury 7 142
Mighty Eight Museum, Savannah, GA 153
Miles M-19 Master II 77
Minneapolis Aquatennial Air Classic Award 37
Mitchell, Billy 23, 156, 157, 168
Mollison, Amy 40–41, 72
Monroe, Marilyn 59
Morrison, James 121, 130
Mountain, Joe 101

National Aeronautic Association 47
National Air and Space Act 142
National Air and Space Administration (NASA) 142, 146
National Air and Space Museum 153
New York Times 117, 119, 122, 137
Newsweek 95
Nichols, Dottie 106–107
Nichols, Ruth 26–27
Nicholson, Mary 75, 77
Night at the Museum: Battle of the Smithsonian 151
Nikumaroro 63
Nixon, Richard 141
Nobel Peace Prize 166
Nolan, Mae Ella 138
Noonan, Fred 62, 160
Northrop 52
Northrop Gamma 41–43, 52
Northrop T-38 37, 146
Nyman, Geri Lamphere 88

Odlum, Bruce 55, 147
Odlum, Floyd 2, 9, 18–19, 30, 32–33, 36, 54–57, 61–62, 69, 123, 132 -133, 136, 139, 147–148, 166–167
Odlum, Stanley 55
O'Hara, Joseph 121–122
Olds, Robert 73, 85–86

P-35 38–39, 43, 52, 168
P-36 168
P-39 Airacobra 103, 106, 108

Index

P-47 Thunderbolt 103, 108, 110, 149, 168
P-51 Mustang 103, 127, 133, 136
P-51B 133
P-63 108
P-80 Shooting Star 134
Parrish, Deanie 131
Pearl Harbor 74, 80, 82, 84
Pegasus (magazine) 111
Pelosi, Nancy 131
Perlich, Robert 44
Pershing, Gen. John J. 164
Pettys, Anna Mae 97
Philippine Merit Medal 162
Phillips, John 139
Pickford, Mary 48
Piper Cobras 131
Pittman, Bessie 9–10, 15, 49
Pittman, Ira 10
Pittman, Mary (Grant) 10
Poole, Barbara 102
Post, Wiley 23, 33
Potsdam Conference 165
Prat and Whitney 52
Princess Martha of Norway 73
PT-17 88
PT-19 88, 108
Pulitzer Air Race 40
Putnam, George 26, 61–64, 160

Quimby, Harriet 25

R-37 108
Raiche, Bessica 25
Ramspeck, Robert 120
Rankin, Jeannette 138
Republic Aviation Corporation 168
Republic F-84 134
Republic Steel 127
Richey, Helen 24, 26–27, 47, 75, 127
Ride, Sally 146
Roach, Eileen 104
Robinson, Marie Michelle 107
Rogers, Edith 82
Rohrer, Alyce Stevens 126
Roosevelt, Eleanor 3, 36, 67, 73
Roosevelt, Franklin 68, 71, 73–74, 165
Roosevelt, Sara 73
Roosevelt Field 31
Roosevelt Flying School 31
Roper, Daniel 63
Royal Air Force 64, 66, 77, 79, 134
Rubinstein, Helen 57
Ryan Flying School 33

Sabre Plaque Award 147
Saund, Dalip Singh 140–141
Scharr, Adela Rick 95, 107, 112, 124

Schneider Air Race 40
Scientific Advisory Group 134
Scott, Blanche 25
Semper Paratus (SPARS) 83
Seversky, Alexander de 38–39 44, 51–52, 167, 169
Seversky Aircraft Corporation 168
Sharp, Evelyn 107
Signal Corps 155
Sinclair, Frank 43
Slone, Geraldine 144
Smith, Fran 106
Smith, Luke 87
Smith, Margaret Chase 138
Smith, Wesley 33, 40
Smithsonian Art Museum 153
Soaring Wings (book) 160
Society of Experimental Test Pilots 37, 147
Spartan 7W 44
Speer, Albert 134
Sperry Gyroscope Company 168
The Spirit of St. Louis 22
Spitfire (British) 75
Sputnik I 142
Stanley, Robert 135
Stars and Stripes 143
Steadman, Bernice Trimble 144
Stewart, Daniel J. 109
Stiller, Ben 151
Stilwell, Joseph 165
Stimson, Henry 118–119
Stine, Betty 106–107
Stinson, Katherine 20, 26
Strok, Mary 107, 113
Strother, Dora Dougherty 1
Stumbough, Gene Nora 144
Sullivan, Madeline 104, 128
Swank, Hillary 151
Sweetwater, Texas 91, 94

Tackaberry, Betty 127
Tamplin, Margaret Chamberlain 103
Thaden, Louis 26–28, 47
Thompson Trophy 24
Thomson-Houston Company 133
Time 112
Towers, John H. 78
towing targets 103–104, 109–110, 123
Truman, Harry S. 166, 169
Tunner, William 85, 87, 163
Turner, Roscoe 23, 42, 47
Tuskegee Airmen 151

UC-18 88
UC-43 88
UC-78 88
United States Air Force Academy 153

Index

U.S. Navy School of Aviation Medicine 144
United World Federalists (UWF) 140

Vale, Scott 59
Victory Through Air Power (book) 168
von Karman, Dr. Theodore 134
von Ohain, Hans J.P. 133
Vorys, John 121–122

War Plans Division of the Army 165
Washington Post 122
Watson, Edwin 73
Weather Wing 105–106
Webb, James E. 143
Wells, Gillis 61
Wendell-Williams 44
West Point (U.S. Military Academy) 155
Whelan, Joan 106
Whittle, Sir Frank 133
Wightman, Hazel 20
Williams, Betty Jane 113
Willis, Helen 20
Wings of Air Medals 37
"Wings to Beauty" 59, 147
Women Accepted for Volunteer Emergency Service (WAVE) 82–83, 94
Women Ferry Pilots 94
Women's Air Derby 40, 48–49
Women's Air Force Ferry Service (WAF) 84–87, 93–95, 98, 100, 103, 106, 109
Women's Airforce Service Pilots (WASPs) 1, 2, 4, 6, 80, 95–96, 98–99, 102, 148, 163, 164, 166, 171; Class 42–1 94; Class 42–2 94; Class 42–3 94; Class 43–1 88, 92–93; Class 43–2 91, 93; Class 44-W-2 96; Museum 153
Women's Army Corps (WAC) 82–85, 92, 94, 118, 162
Women's Auxiliary Air Corps (U.S.) 73
Women's Auxiliary Air Force (British) 66
Women's Auxiliary Army Corps (WAAC) 82, 161
Women's Auxiliary Ferry Squadron (WAFS) 103, 106, 109, 163
Women's Auxiliary Territorial Service 66
Women's Flying Training Detachment (WFTD) 90–91, 95
Women's Flying Training Program 86
Women's National Aeronautical Association 37
Women's National Air Meet 48
Women's Voluntary Services 66
Wood, Anne 75, 79
Wood, Winifred 109, 127
Wrath in Burma (book) 140
Wright Brothers 21, 25, 155, 157

Yalta Conference 165
Yeager, Chuck 3–4, 55, 132, 133, 136–138, 166
Yeager, Glennis 166
Yount, Barton K. 74, 86–87, 92, 169, 171
YP-59 135

www.ingramcontent.com/pod-product-compliance
Ingram Content Group UK Ltd.
Pitfield, Milton Keynes, MK11 3LW, UK
UKHW041958140426
5217IPUK00015B/865